Springer Theses

Recognizing Outstanding Ph.D. Research

For further volumes:
http://www.springer.com/series/8790

Aims and Scope

The series "Springer Theses" brings together a selection of the very best Ph.D. theses from around the world and across the physical sciences. Nominated and endorsed by two recognized specialists, each published volume has been selected for its scientific excellence and the high impact of its contents for the pertinent field of research. For greater accessibility to non-specialists, the published versions include an extended introduction, as well as a foreword by the student's supervisor explaining the special relevance of the work for the field. As a whole, the series will provide a valuable resource both for newcomers to the research fields described, and for other scientists seeking detailed background information on special questions. Finally, it provides an accredited documentation of the valuable contributions made by today's younger generation of scientists.

Theses are accepted into the series by invited nomination only and must fulfill all of the following criteria

- They must be written in good English.
- The topic should fall within the confines of Chemistry, Physics and related interdisciplinary fields such as Materials, Nanoscience, Chemical Engineering, Complex Systems and Biophysics.
- The work reported in the thesis must represent a significant scientific advance.
- If the thesis includes previously published material, permission to reproduce this must be gained from the respective copyright holder.
- They must have been examined and passed during the 12 months prior to nomination.
- Each thesis should include a foreword by the supervisor outlining the significance of its content.
- The theses should have a clearly defined structure including an introduction accessible to scientists not expert in that particular field.

Zaozao Qiu

Late Transition Metal-Carboryne Complexes

Synthesis, Structure, Bonding, and Reaction with Alkenes and Alkynes

Doctoral Thesis accepted by
The Chinese University of Hong Kong, China

Author
Dr. Zaozao Qiu
Department of Chemistry
The Chinese University of Hong Kong
Shatin, N. T., Hong Kong
People's Republic of China
e-mail: zaozaoqiu@cuhk.edu.hk

Supervisor
Prof. Zuowei Xie
Department of Chemistry
The Chinese University of Hong Kong
Shatin, N. T., Hong Kong
People's Republic of China

ISSN 2190-5053
ISBN 978-3-642-24360-8
DOI 10.1007/978-3-642-24361-5
Springer Heidelberg Dordrecht London New York

e-ISSN 2190-5061
e-ISBN 978-3-642-24361-5

Library of Congress Control Number: 2011938998

© Springer-Verlag Berlin Heidelberg 2012
This work is subject to copyright. All rights are reserved, whether the whole or part of the material is concerned, specifically the rights of translation, reprinting, reuse of illustrations, recitation, broadcasting, reproduction on microfilm or in any other way, and storage in data banks. Duplication of this publication or parts thereof is permitted only under the provisions of the German Copyright Law of September 9, 1965, in its current version, and permission for use must always be obtained from Springer. Violations are liable to prosecution under the German Copyright Law.
The use of general descriptive names, registered names, trademarks, etc. in this publication does not imply, even in the absence of a specific statement, that such names are exempt from the relevant protective laws and regulations and therefore free for general use.

Printed on acid-free paper

Springer is part of Springer Science+Business Media (www.springer.com)

Parts of this thesis have been published in the following journal articles

Zaozao Qiu and Zuowei Xie. "Nickel-Mediated Coupling Reaction of Carboryne with Alkenes: New Synthetic Route to Alkenylcarboranes" *Angew. Chem. Int. Ed.* **2008**, *47*, 6572–6575. *Reproduced with permission*

Zaozao Qiu and Zuowei Xie. "Nickel-Mediated Three-Component Cycloaddition Reaction of Carboryne, Alkenes, and Alkynes" *J. Am. Chem. Soc.* **2009**, *131*, 2084–2085. *Reproduced with permission*

Zaozao Qiu and Zuowei Xie. "Nickel-Catalyzed Three-Component [2+2+2] Cycloaddition Reaction of Arynes, Alkenes, and Alkynes" Angew. *Chem. Int. Ed.* **2009**, *48*, 5729–5732. *Reproduced with permission*

Zaozao Qiu, Sunewang R. Wang, and Zuowei Xie. "Nickel-Catalyzed Regioselective [2+2+2] Cycloaddition of Carboryne with Alkynes" *Angew. Chem. Int. Ed.* **2010**, *49*, 4649–4652. *Reproduced with permission*

Zaozao Qiu and Zuowei Xie. "Palladium/nickel-cocatalyzed cycloaddition of 1,3-dehydro-*o*-carborane with alkynes. Facile synthesis of C,B-substituted carboranes" *J. Am. Chem. Soc.* **2010**, *132*, 16085–16093. *Reproduced with permission*

Zaozao Qiu, Liang Deng, Hoi-Shan Chan, and Zuowei Xie. "Synthesis and structural characterization of group 10 metal−carboryne complexes" *Organometallics* **2010**, *9*, 4541–4547. *Reproduced with permission*

Supervisor's Foreword

Carboranes are a class of boron hydride clusters in which one or more of the BH vertices are replaced by CH units. Their unique properties such as outstanding thermal and chemical stability, three-dimensional structure, σ-aromaticity and special electronic properties have made them as useful building blocks for their use in luminescent materials, polymers, dendrimers, liquid crystals and nonlinear optics, as potent BNCT (boron neutron capture treatment) agents in medicine, and as versatile ligands in organometallic/coordination chemistry. On the other hand, excellent chemical stability of carboranes makes their derivatization very difficult, which has restricted the applications within a narrow scope.

Dr. Qiu takes the challenge and develops new methodologies for the functionalization of carboranes via the metal-carboryne (carboryne=1,2-dehydro-o-carborane) intermediates. She has prepared a series of nickel-carboryne complexes and studied their structures. The results suggest that the bonding interactions between the nickel atom and the carboryne ligand are best described as a resonance hybrid of both the Ni–C σ and Ni–C π bonds, similar to those described for metal-benzyne complexes. Subsequently, Dr. Qiu investigates the reaction chemistry of the nickel-carboryne complexes and obtains many important results. For example, $(\eta^2\text{-}C_2B_{10}H_{10})Ni(PPh_3)_2$ can undergo regioselective [2+2+2] cycloaddition reactions with 2 equiv of alkyne to afford benzocarboranes, react with 1 equiv of alkene to generate alkenylcarborane coupling products, and also undergo a three-component [2+2+2] cyclotrimerization with 1 equiv of activated alkene and 1 equiv of alkyne to give dihydrobenzocarboranes. The reaction of carboryne with alkynes can be catalyzed by Ni species. Accordingly, a Pd/Ni-co-catalyzed [2+2+2] cycloaddition reaction of 1,3-dehydro-o-carborane with 2 equiv of alkyne has also been developed, leading to the efficient formation of C,B-substituted benzocarboranes in a single process.

Dr. Qiu's work breaks a new ground in metal-carboryne chemistry. The results detailed in this thesis will further that effort by providing easy access to a wide range of functionalized carborane derivatives.

Hong Kong, September 2011   Zuowei Xie

Acknowledgments

I would like to give my sincere thanks to my supervisor, Professor Zuowei Xie, for giving the opportunity to work in his research group and study the fascinating chemistry of metal-carboryne. I am grateful for his guidance, encouragement, and help during my study in the past four years.

Great thanks are also given to Professor Kazushi Mashima for his advice and suggestions in my three months' study in his laboratory in Osaka University.

I am also grateful to my groupmates Dr. Liang Deng, Dr. Hao Shen, Dr. Shikuo Ren, Ms. Dongmei Liu, Mr. Jian Zhang, Ms. Mei-Mei Sit, Mr. Xiaodu Fu, Ms. Jingnan Chen, Mr. Fangrui Zheng, Ms. Jingying Yang, Mr. Sunewang R. Wang, Mr. Yang Wang, and Mr. Xiao He for their help and cooperation.

I would like to thank Ms. Hoi-Shan Chan for the determination of single-crystal X-ray structures, Ms. Hau-Yan Ng for mass spectra measurement, and Mr. Chun-Wah Lin for GC-MS measurement. I also thank the staff and technicians in the Department of Chemistry and Graduate School for their help and supports during the course of my study.

I am greatly indebted to The Chinese University of Hong Kong for the award of a Postgraduate Studentship and the Hong Kong Research Grants Council for the financial support.

Finally, I would like to dedicate this thesis to my parents for their unconditional love, care, understanding, and support throughout these years.

Contents

1 Introduction 1
  1.1 Synthesis of C-Substituted Carboranes 2
      1.1.1 Reaction of Decaborane with Acetylenes 2
      1.1.2 Reaction of Lithiocarborane with Electrophiles 3
      1.1.3 Reaction of Carboranylcopper 5
      1.1.4 Other Methods 7
  1.2 Synthesis of B-Substituted Carboranes 8
      1.2.1 Reaction of Decaborane with Acetylenes 8
      1.2.2 Electrophilic Substitution of Carboranes 8
      1.2.3 Reaction of Dicarbollide Ion ($C_2B_9H_{11}{}^{2-}$)
           with Dihalobarane 9
      1.2.4 Transition Metal-Catalyzed Coupling Reaction 10
      1.2.5 Carbene Reaction 12
      1.2.6 Other Methods 12
  1.3 1,2-*o*-Carboryne 13
  1.4 Transition Metal-1,2-*o*-Carboryne Complexes 16
      1.4.1 Synthesis of Transition Metal-1,2-*o*-Carboryne
           Complexes 17
      1.4.2 Reactivity of Zr-1,2-*o*-Carboryne Complexes 20
      1.4.3 Reactivity of Ni-1,2-*o*-Carboryne Complexes 24
  1.5 Our Objectives 25
  References 26

2 Nickel-1,2-*o*-Carboryne Complexes 31
  2.1 Introduction 31
  2.2 Synthesis and Structure of B-Substituted
      Nickel-1,2-*o*-Carboryne Complexes 32
  2.3 Summary 36
  References 36

xii Contents

3 Nickel-Mediated Coupling Reaction of 1,2-*o*-Carboryne with Alkenes 39
   3.1 Introduction 39
   3.2 Results and Discussion 40
   3.3 Summary 53
   References 54

4 Nickel-Mediated/Catalyzed Three-Component Cycloaddition Reaction of 1,2-*o*-Carboryne/Arynes, Alkenes, and Alkynes 55
   4.1 Introduction 55
   4.2 Results and Discussion 56
   4.3 Summary 68
   References 68

5 Nickel-Catalyzed [2+2+2] Cycloaddition of 1,2-*o*-Carboryne with Alkynes 71
   5.1 Introduction 71
   5.2 Results and Discussion 71
   5.3 Summary 82
   References 82

6 Palladium/Nickel-Cocatalyzed [2+2+2] Cycloaddition of 1,3-*o*-Carboryne with Alkynes 85
   6.1 Introduction 85
   6.2 Results and Discussion 86
      6.2.1 Synthesis of 1,3-*o*-Carboryne Precursor 86
      6.2.2 Reaction of 1,3-*o*-Carboryne Precursor 87
      6.2.3 Metal-Catalyzed [2+2+2] Cycloaddition of 1,3-*o*-Carboryne with Alkynes 88
      6.2.4 Proposed Mechanism 92
   6.3 Summary 95
   References 95

7 Conclusion 99

8 Experimental Section 101
   References 126

Appendix 129

Abbreviation

bipy	bipyridine
^{n}Bu (or *n*-Bu)	*n*-butyl
^{n}BuLi	*n*-butyl lithium
^{t}Bu (or *t*-Bu)	*t*-butyl
cat.	catalyst
cod	cyclooctadiene
Cp	cyclopentadienyl
Cy	cyclohexyl
d	doublets
dba	dibenzylideneacetone
dcpe	1,2-bis(dicyclohexylphosphino)ethane
dd	doublet of doublets
DME	dimethoxyethane
dppe	1,2-bis(diphenylphosphino)ethane
dppen	*cis*-1,2-bis(diphenylphosphino)ethene
dppp	1,3-bis(diphenylphosphino)propane
eq. (or equiv)	equivalent
Et	ethyl
Et_2O	diethyl ether
IR	infrared spectroscopy
L	ligand
m	multiplet
M	metal
Me	methyl
Me_2Im	1,3-dimethylimidazol-2-ylidene
MS	mass spectroscopy
NMR	nuclear magnetic resonance spectroscopy
Ph	phenyl
^{n}Pr (or *n*-Pr)	propyl
^{i}Pr (or *i*-Pr)	isopropyl
pyr	pyridine

r.t. (or RT)	room temperature
s	singlet
t	triplet
TBAF	tetrabutylammonium floride
TBDMS	*t*-butyldimethylsilyl
THF	tetrahydrofuran
TMS	trimethylsilyl
Ts	Toluenesulfonyl
xs	excess

List of Compounds

Compd No.	Compound Formula	Page No.
II-1	$(\eta^2\text{-}9\text{-I-}1,2\text{-}C_2B_{10}H_9)Ni(PPh_3)_2$	32
II-2	$(\eta^2\text{-}9,12\text{-I}_2\text{-}1,2\text{-}C_2B_{10}H_8)Ni(PPh_3)_2$	32
II-3	$(\eta^2\text{-}3\text{-Br-}1,2\text{-}C_2B_{10}H_9)Ni(PMe_3)_2$	32
II-4	$(\eta^2\text{-}3\text{-}C_6H_5\text{-}1,2\text{-}C_2B_{10}H_9)Ni(PMe_3)_2$	32
II-5	$(\eta^2\text{-}3\text{-}C_6H_5\text{-}1,2\text{-}C_2B_{10}H_9)Ni(PPh_3)_2$	32
II-6	$(\eta^2\text{-}4,5,7,8,9,10,11,12\text{-Me}_8\text{-}C_2B_{10}H_2)Ni(PMe_3)_2$	33
III-3a	*trans*-1-(HC=CHPh)-1,2-$C_2B_{10}H_{11}$	103
[D$_3$]-III-3a	*trans*-1-[DC=CD(Ph)]-2-D-1,2-$C_2B_{10}H_{11}$	104
III-3b	*trans*-1-{HC=CH[(4′-CH$_3$)C$_6$H$_4$]}-1,2-$C_2B_{10}H_{11}$	104
III-3c	*trans*-1-{HC=CH[(4′-CF$_3$)C$_6$H$_4$]}-1,2-$C_2B_{10}H_{11}$	104
III-3d	*trans*-1-{HC=CH[(3′-CF$_3$)C$_6$H$_4$]}-1,2-$C_2B_{10}H_{11}$	104
III-3e	*trans*-1-{HC=CH[3′,4′,5′-(OMe)$_3$C$_6$H$_2$]}-1,2-$C_2B_{10}H_{11}$	104
III-4f	1-[H$_2$CC(Ph)=CH$_2$]-1,2-$C_2B_{10}H_{11}$	104
III-3g	1-[HC=C(Ph)$_2$]-1,2-$C_2B_{10}H_{11}$	104
III-3h	*trans*-1-[HC=CH(SiMe$_3$)]-1,2-$C_2B_{10}H_{11}$	105
III-4i	1-[H$_2$CC=CH(CH$_2$)$_3$CH$_2$]-1,2-$C_2B_{10}H_{11}$	105
III-4j	*trans*-1-[H$_2$CCH=CH(CH$_2$)$_3$]-1,2-$C_2B_{10}H_{11}$	105
III-5j	*cis*-1-[H$_2$CCH=CH(CH$_2$)$_3$]-1,2-$C_2B_{10}H_{11}$	105
III-4k	1-[HCC=CH(CH$_2$)$_2$CH$_2$]-1,2-$C_2B_{10}H_{11}$	105
III-5l	1-bicyclo[2.2.1]hept-2-yl-1,2-carborane	105
III-3m	1-(1H-inden-2-yl)-1,2-carborane	106
III-5m	1-(2,3-dihydro-1H-inden-2-yl)-1,2-carborane	106
III-3n	*trans*-1-[HC=CH(O^{n}Bu)]-1,2-$C_2B_{10}H_{11}$	106
III-5n	1-[HC(Me)(O^{n}Bu)]-1,2-$C_2B_{10}H_{11}$	106
III-4o	1-[HCC=CH(CH$_2$)$_2$O]-1,2-$C_2B_{10}H_{11}$	106

(continued)

xvi — List of Compounds

(continued)

Compd No.	Compound Formula	Page No.
III-5p	1-[$CH_2CH_2(CO_2Me)$]-1,2-$C_2B_{10}H_{11}$	106
[D_2]-III-5p	1-[$CH_2CH(D)(CO_2Me)$]-2-D-1,2-$C_2B_{10}H_{11}$	107
III-6p	1-[$CH_2CH(CO_2Me)CH_2CH_2CO_2Me$]-1,2-$C_2B_{10}H_{11}$	107
III-5q	1-[$CH_2CH_2(o-C_5H_4N)$]-1,2-$C_2B_{10}H_{11}$	107
III-7q	1,2-[$CH_2CH_2(o-C_5H_4N)$]$_2$-1,2-$C_2B_{10}H_{10}$	107
III-8q	*trans*-1-[CH=CH($o-C_5H_4N$)]-2-[$CH_2CH_2(o-C_5H_4N)$]-1,2-$C_2B_{10}H_{10}$	107
III-9p	[2-$CH_2CH(CO_2Me)$-1,2-$C_2B_{10}H_{10}$]Ni(PPh_3)	108
III-9q	[{[2-$CH_2CH(o-C_5H_4N)$-1,2-$C_2B_{10}H_{10}$]Ni}$_3$(μ_3-Cl)][Li(DME)$_3$]	108
IV-1a	1,2-[EtC=C(Et)CH($o-C_5H_4N$)CH_2]-1,2-$C_2B_{10}H_{10}$	109
IV-1b	1,2-[^nBuC=C(^nBu)CH($o-C_5H_4N$)CH_2]-1,2-$C_2B_{10}H_{10}$	109
IV-1c	1,2-[^iPrC=C(Me)CH($o-C_5H_4N$)CH_2]-1,2-$C_2B_{10}H_{10}$	109
IV-1c'	1,2-[MeC=C(^iPr)CH($o-C_5H_4N$)CH_2]-1,2-$C_2B_{10}H_{10}$	109
IV-1d	1,2-[PhC=C(Me)CH($o-C_5H_4N$)CH_2]-1,2-$C_2B_{10}H_{10}$	110
IV-1e	1,2-[(4'-Me-C_6H_4)C=C(Me)CH($o-C_5H_4N$)CH_2]-1,2-$C_2B_{10}H_{10}$	110
IV-1f	1,2-[PhC=C(Et)CH($o-C_5H_4N$)CH_2]-1,2-$C_2B_{10}H_{10}$	110
IV-1g	1,2-[PhC=C(^nBu)CH($o-C_5H_4N$)CH_2]-1,2-$C_2B_{10}H_{10}$	110
IV-1h	1,2-[PhC=C(CH_2CH=CH_2)CH($o-C_5H_4N$)CH_2]-1,2-$C_2B_{10}H_{10}$	110
IV-1i	1,2-[EtC=C(Et)CH(CO_2Me)CH_2]-1,2-$C_2B_{10}H_{10}$	111
IV-1j	1,2-[^nPrC=C(^nPr)CH(CO_2Me)CH_2]-1,2-$C_2B_{10}H_{10}$	111
IV-1k	1,2-[^nBuC=C(^nBu)CH(CO_2Me)CH_2]-1,2-$C_2B_{10}H_{10}$	111
IV-5a	1,2-[PhC=C(Ph)CH(CO_2Me)CH_2]C_6H_4	111
IV-5b	1,2-[PhC=C(Ph)CH(CO_2^nBu)CH_2]C_6H_4	112
IV-5c	1,2-[PhC=C(Ph)CH(CO_2^iBu)CH_2]C_6H_4	112
IV-5d	1,2-{PhC=C(Ph)[CHC(=O)Me]CH_2}C_6H_4	112
IV-5e	1,2-[PhC=C(Ph)CH(CN)CH_2]C_6H_4	112
IV-5f	1,2-(OCH_2O)-4,5-[PhC=C(Ph)CH(CO_2Me)CH_2]C_6H_2	112
IV-5g	4,5-(CH_2)$_3$-1,2-[PhC=C(Ph)CH(CO_2Me)CH_2]C_6H_2	113
IV-5h	4-Me-1,2-[PhC=C(Ph)CH(CO_2Me)CH_2]C_6H_3	113
IV-5'h	5-Me-1,2-[PhC=C(Ph)CH(CO_2Me)CH_2]C_6H_3	113
IV-5i	1,2-[MeC=C(Ph)CH(CO_2Me)CH_2]C_6H_4	113
IV-5j	1,2-[EtC=C(Ph)CH(CO_2Me)CH_2]C_6H_4	113
IV-5k	1,2-[^nBuC=C(Ph)CH(CO_2Me)CH_2]C_6H_4	113
IV-5l	1,2-[C(CH_2OMe)=C(Ph)CH(CO_2Me)CH_2]C_6H_4	114
IV-5m	1,2-[C(CH_2CH=CH_2)=C(Ph)CH(CO_2Me)CH_2]C_6H_4	114
IV-5n	1,2-{C[(CH_2)$_3$CN]=C(Ph)CH(CO_2Me)CH_2}C_6H_4	114
IV-5o	1,2-[MeC=C(4'-Me-C_6H_4)CH(CO_2Me)CH_2]C_6H_4	114
IV-5p	1,2-[MeC=C(CO_2Me)CH(CO_2Me)CH_2]C_6H_4	114
IV-5q	1,2-[EtC=C(Et)CH(CO_2Me)CH_2]C_6H_4	115
IV-5r	1,2-[^nBuC=C(^nBu)CH(CO_2Me)CH_2]C_6H_4	115
IV-5s	1,2-[^iPrC=C(Me)CH(CO_2Me)CH_2]C_6H_4	115
IV-5s'	1,2-[MeC=C(^iPr)CH(CO_2Me)CH_2]C_6H_4	115
V-1a	1,2-[EtC=C(Et)C(Et)=CEt]-1,2-$C_2B_{10}H_{10}$	116
V-1b	3-Cl-1,2-[EtC=C(Et)C(Et)=CEt]-1,2-$C_2B_{10}H_9$	117
V-1c	3-Ph-1,2-[EtC=C(Et)C(Et)=CEt]-1,2-$C_2B_{10}H_9$	117

(continued)

List of Compounds xvii

(continued)

Compd No.	Compound Formula	Page No.
V-1d	1,2-[^{n}PrC=C(^{n}Pr)C(^{n}Pr)=C^{n}Pr]-1,2-C$_2$B$_{10}$H$_{10}$	117
V-1e	1,2-[^{n}BuC=C(^{n}Bu)C(^{n}Bu)=C^{n}Bu]-1,2-C$_2$B$_{10}$H$_{10}$	117
V-1f	1,2-[PhC=C(Ph)C(Ph)=CPh]-1,2-C$_2$B$_{10}$H$_{10}$	117
V-1g	1,2-[C(CH$_2$OMe)=C(CH$_2$OMe)C(CH$_2$OMe)=C(CH$_2$OMe)]-1,2-C$_2$B$_{10}$H$_{10}$	117
V-1h	1,2-[MeC=C(^{i}Pr)C(Me)=C^{i}Pr]-1,2-C$_2$B$_{10}$H$_{10}$	118
V-1'h	1,2-[MeC=C(^{i}Pr)C(^{i}Pr)=CMe]-1,2-C$_2$B$_{10}$H$_{10}$	118
V-1i	1,2-[MeC=C(Ph)C(Me)=CPh]-1,2-C$_2$B$_{10}$H$_{10}$	118
V-1j	1,2-[MeC=C(4'-Me-C$_6$H$_4$)C(Me)=C(4'-Me-C$_6$H$_4$)]-1,2-C$_2$B$_{10}$H$_{10}$	118
V-1k	1,2-[MeC=C(4'-CF$_3$-C$_6$H$_4$)C(Me)=C(4'-CF$_3$-C$_6$H$_4$)]-1,2-C$_2$B$_{10}$H$_{10}$	118
V-1l	1,2-[EtC=C(Ph)C(Et)=CPh]-1,2-C$_2$B$_{10}$H$_{10}$	118
V-1m	1,2-[^{n}BuC=C(Ph)C(^{n}Bu)=CPh]-1,2-C$_2$B$_{10}$H$_{10}$	118
V-1n	1,2-[C(C≡CPh)=C(Ph)C(C≡CPh)=CPh]-1,2-C$_2$B$_{10}$H$_{10}$	119
V-1o	1,2-[C(CH$_2$OMe)=C(Ph)C(CH$_2$OMe)=CPh]-1,2-C$_2$B$_{10}$H$_{10}$	119
V-1'o	1,2-[PhC=C(CH$_2$OMe)C(CH$_2$OMe)=CPh]-1,2-C$_2$B$_{10}$H$_{10}$	119
V-2a	1,2-(2-methyl-2,5-cyclohexadiene-1,4-diyl)-o-carborane	115
V-2b	1,2-(1-methyl-2,5-cyclohexadiene-1,4-diyl)-o-carborane	116
V-3	1-[C(Et)=C=CH(Me)]-1,2-C$_2$B$_{10}$H$_{11}$	116
V-6o	1-[C(CH$_2$OMe)=CH(Ph)]-1,2-C$_2$B$_{10}$H$_{11}$	119
V-6'o	1-[C(Ph)=CH(CH$_2$OMe)]-1,2-C$_2$B$_{10}$H$_{11}$	119
V-8a	1,2-[MeC=C̅-(CH$_2$)$_3$-C̅=CMe]-1,2-C$_2$B$_{10}$H$_{10}$	119
V-8b	1,2-[MeC=C̅-(CH$_2$)$_4$-C̅=CMe]-1,2-C$_2$B$_{10}$H$_{10}$	120
V-8c	1,2-[MeC=C̅-(CH$_2$)$_5$-C̅=CMe]-1,2-C$_2$B$_{10}$H$_{10}$	120
V-9	[{[2-C(^{n}Bu)=C(o-C$_5$H$_4$N)-1,2-C$_2$B$_{10}$H$_{10}$]Ni}$_2$(μ_2-Cl)] [Li(THF)$_4$]	120
VI-1a	3-I-1,2-C$_2$B$_{10}$H$_{11}$	120
VI-1b	3-I-1-Me-1,2-C$_2$B$_{10}$H$_{10}$	121
VI-1c	3-I-1-Ph-1,2-C$_2$B$_{10}$H$_{10}$	121
VI-1d	1-^{n}Bu-3-I-1,2-C$_2$B$_{10}$H$_{10}$	121
VI-1e	3-I-1-TMS-1,2-C$_2$B$_{10}$H$_{10}$	122
VI-1f	3-I-1-(CH$_2$CH$_2$OCH$_3$)-1,2-C$_2$B$_{10}$H$_{10}$	122
VI-1g	3-I-1-[CH$_2$CH$_2$N(CH$_3$)$_2$]-1,2-C$_2$B$_{10}$H$_{10}$	122
VI-4a	1,3-[EtC=C(Et)C(Et)=CEt]-1,2-C$_2$B$_{10}$H$_{10}$	122
VI-4b	2-Me-1,3-[EtC=C(Et)C(Et)=CEt]-1,2-C$_2$B$_{10}$H$_9$	123
VI-4c	2-^{n}Bu-1,3-[EtC=C(Et)C(Et)=CEt]-1,2-C$_2$B$_{10}$H$_9$	123
VI-4d	2-TMS-1,3-[EtC=C(Et)C(Et)=CEt]-1,2-C$_2$B$_{10}$H$_9$	123
VI-4e	2-Ph-1,3-[EtC=C(Et)C(Et)=CEt]-1,2-C$_2$B$_{10}$H$_9$	123
VI-4f	2-(CH$_2$CH$_2$OCH$_3$)-1,3-[EtC=C(Et)C(Et)=CEt]-1,2-C$_2$B$_{10}$H$_9$	123
VI-4g	2-[CH$_2$CH$_2$N(CH$_3$)$_2$]-1,3-[EtC=C(Et)C(Et)=CEt]-1,2-C$_2$B$_{10}$H$_9$	124
VI-4h	2-Me-1,3-[^{n}PrC=C(^{n}Pr)C(^{n}Pr)=C^{n}Pr]-1,2-C$_2$B$_{10}$H$_9$	124
VI-4i	2-Me-1,3-[^{n}BuC=C(^{n}Bu)C(^{n}Bu)=C^{n}Bu]-1,2-C$_2$B$_{10}$H$_9$	124
VI-4j	2-Me-1,3-[PhC=C(Ph)C(Ph)=CPh]-1,2-C$_2$B$_{10}$H$_9$	124
VI-4k	2-Me-1,3-[C(4'-Me-C$_6$H$_4$)=C(4'-Me-C$_6$H$_4$)C(4'-Me-C$_6$H$_4$)=C(4'-Me-C$_6$H$_4$)]-1,2-C$_2$B$_{10}$H$_9$	124

(continued)

xviii
List of Compounds

(continued)

Compd No.	Compound Formula	Page No.
VI-4l	2-Me-1,3-[PhC=C(Me)C(Ph)=CMe]-1,2-$C_2B_{10}H_9$	125
VI-5l	2-Me-1,3-[MeC=C(Ph)C(Ph)=CMe]-1,2-$C_2B_{10}H_9$	
VI-4m	2-Me-1,3-[PhC=C(Et)C(Ph)=CEt]-1,2-$C_2B_{10}H_9$	125
VI-5m	2-Me-1,3-[EtC=C(Ph)C(Ph)=CEt]-1,2-$C_2B_{10}H_9$	
VI-7a	2-Me-1,3-[MeC=$\overline{\text{C-(CH}_2)_3\text{-C}}$=CMe]-1,2-$C_2B_{10}H_9$	125
VI-7b	2-Me-1,3-[MeC=$\overline{\text{C-(CH}_2)_4\text{-C}}$=CMe]-1,2-$C_2B_{10}H_9$	126
VI-7c	2-Me-1,3-[MeC=$\overline{\text{C-(CH}_2)_5\text{-C}}$=CMe]-1,2-$C_2B_{10}H_9$	126

Chapter 1
Introduction

Carboranes (dicarba-*closo*-dodecaboranes), members of the class of carbon-containing boron clusters, have characteristic properties such as spherical geometry, remarkable thermal and chemical stability, and a hydrophobic molecular surface. Medical applications of the carboranes have been mainly in the field of boron neutron capture therapy (BNCT) of cancer, utilizing the high boron content of the carboranes. (For recent reviews on medicinal applications, see Ref.[1–4])

The synthesis and properties of icosohedral carboranes were first reported in 1963 [5–8], which have been the most extensively investigated of all known carboranes during the last 40 years [9–20]. *o*-Carborane was obtained by the reaction of acetylene with complexes prepared from decaborane and Lewis base such as acetonitrile, alkylamines, and alkyl sulfides (Scheme 1.1) [5–8, 21, 22].

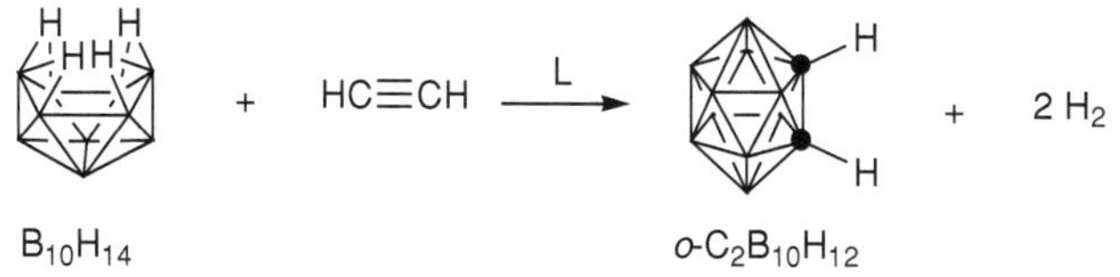

Scheme 1.1 Synthesis of *o*-carborane by the reaction of decaborane with acetylene

Unlike the starting boron hydride, *o*-carborane is stable in the presence of oxidizing agents, alcohols, and strong acids and exhibits phenomenal thermal stability in temperatures up to 400 °C. Under inert atmosphere, it rearranges to *m*-carborane between 400 and 500 °C. This latter compound isomerizes to *p*-carborane between 600 and 700 °C (Scheme 1.2) [9–15].

One of the most important features of a carborane system is its ability to enter into substitution reactions at both the carbon and boron atoms without degradation of the cage. The stability of the carborane cage is demonstrated under many reaction conditions used to prepare a wide range of C- and B-carborane derivatives [16].

Z. Qiu, *Late Transition Metal-Carboryne Complexes*, Springer Theses, DOI: 10.1007/978-3-642-24361-5_1, © Springer-Verlag Berlin Heidelberg 2012

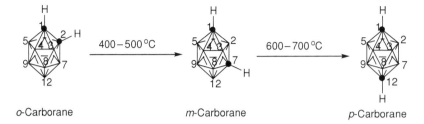

o-Carborane   *m*-Carborane   *p*-Carborane

Scheme 1.2 Rearrangement of carboranes

1.1 Synthesis of C-Substituted Carboranes

During the past decades investigations in the field of C-substituted carboranes were directed at improving the synthetic methods for the preparation of organic and organometallic carboranyl compounds used in the production of polymeric materials as well as in biological and medical investigations. There are two conventional synthetic methods leading to carbon-substituted carboranes: the reaction of substituted acetylenes with decaborane and electrophilic substitutions of lithiocarboranes.

1.1.1 Reaction of Decaborane with Acetylenes

The reaction of decaborane with acetylenes in the presence of Lewis bases is a general method for carborane synthesis. The typical yields of this method for terminal alkynes range from 6 to 75%, whereas much lower yields are given or even no reaction takes place for many internal alkynes [5–8, 21, 22]. Recently, Sneddon and co-workers reported an improved method for the synthesis of 1,2-disubstituted *o*-carboranes by direct reaction of $B_{10}H_{14}$ or 6-R-$B_{10}H_{13}$ with terminal or internal alkynes in ionic liquid in high yields (Scheme 1.3) [23, 24].

Scheme 1.3 Synthesis of *o*-carboranes in ionic liquid

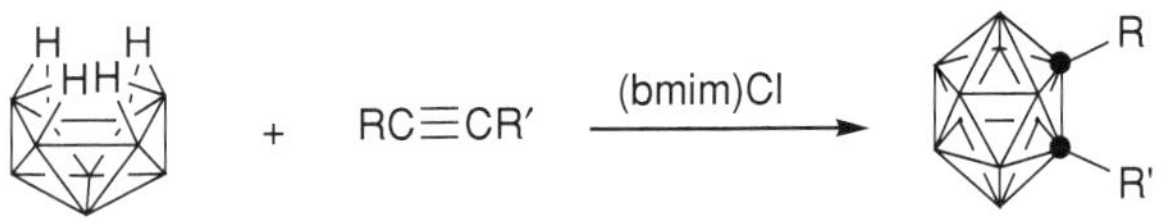

R = R' = Et, Ph; R = Hexyl, R' = H; R = CH$_2$OH, R' = Me
bmim = 1-butyl-3-methylimidiazolium

1.1.2 Reaction of Lithiocarborane with Electrophiles

The strong electron-withdrawing character of the *o*-carborane unit facilitates the metalation of the carborane CH group. The mildly acidic C–H bonds in *o*-carborane (with the *p*Ka value of ~23) react easily with strong bases such as ^{n}BuLi and PhLi to form C-monolithio-*o*-carborane or C,C'-dilithio-*o*-carborane, which can be converted into the corresponding mono- or di-organosubstituted products (Scheme 1.4) [25, 26].

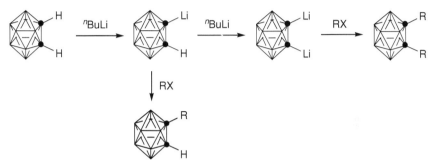

Scheme 1.4 Synthesis of C-substituted carboranes from C-monolithio- and C,C'-dilithio-*o*-caborane

The synthesis of mono-C-substituted *o*-carboranes is not straightforward due to the disproportionation of monolithio-*o*-carborane, which leads to the undesired di-C-substituted products (Scheme 1.5) [26].

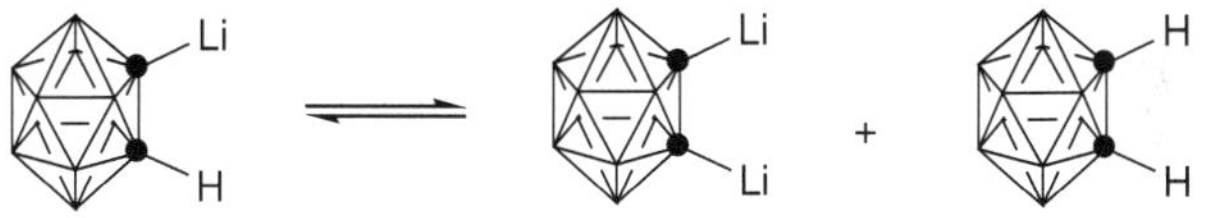

Scheme 1.5 Disproportionation of monolithio-*o*-carborane

Hawthorne developed an effective method to prepare monosubstituted *o*-carboranes using *tert*-butyldimethylsilyl (TBDMS) as a protecting group, as the reactions between mono- or dilithio-*o*-carborane with TBDMSCl only give monosubstituted product and the TBDMS group can be easily removed by TBAF (^{n}Bu$_4$NF) (Scheme 1.6) [27]. The easy functionalization of the cage CH vertices results in the emergence of numerous carborane derivatives, which makes the further application possible [9].

Scheme 1.6 Synthesis of mono-C-substituted carboranes using TBDMS as a protecting group

The elimination of p-toluenesulfonic acid from the corresponding tosylate gives 1,2-ethano-o-carborane in presence of ^{n}BuLi up to 40% yield with the molar ratio of 1,2-ethano-o-carborane to 1-vinyl-o-carborane as high as 99/1 (Scheme 1.7) [28].

Scheme 1.7 Synthesis of 1,2-ethano-o-carborane

Carboranophanes, m-carboranes bridged by a single all-carbon or carbon and sulfur bridge were also synthesized by the action of lithiocarborane on S_8 followed by an alkylation-oxidation-pyrolysis route (Scheme 1.8) [29].

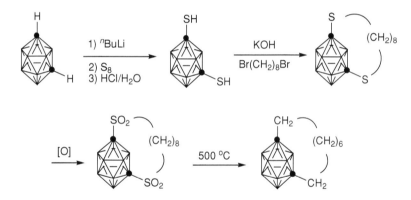

Scheme 1.8 Synthesis of bridged m-carboranes from 1,7-bis(mercapto)-m-carborane

Dilithio-p-carborane can be converted to 1,12-bis(hydroxycarbonyl)-p-carborane in almost quantitative yield by reaction with carbon dioxide followed by acidification (Scheme 1.9) [30].

Scheme 1.9 Synthesis of C-carboxylcarborane

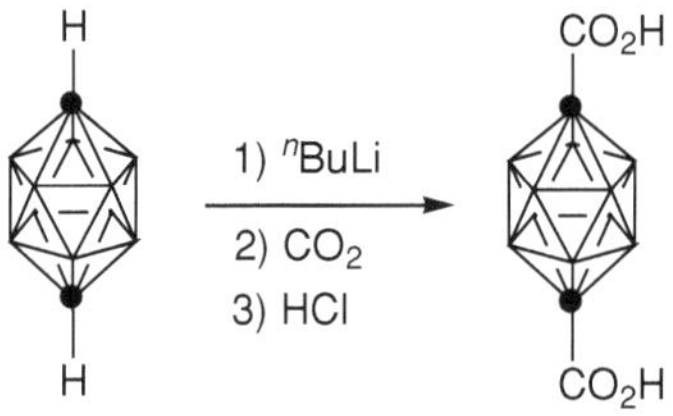

1.1 Synthesis of C-Substituted Carboranes

Single-step preparation of C-formyl derivatives directly from *o*-, *m*-, and *p*-carboranes was reported by Dozzo et al. in 2005 (Scheme 1.10) [31].

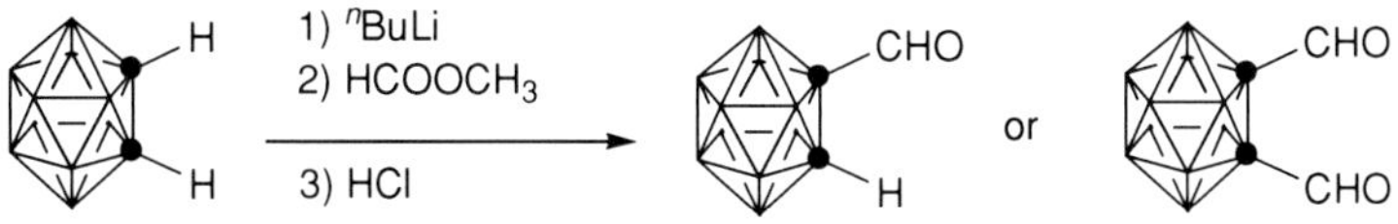

Scheme 1.10 Synthesis of C-formylcarboranes

C-hydroxylcarboranes and C,C-dihydroxylcarboranes can be synthesized by the reaction of lithiocarborane and trimethylborate, followed by oxidation with hydrogen peroxide in the presence of acetic acid through a one-pot procedure (Scheme 1.11) [32].

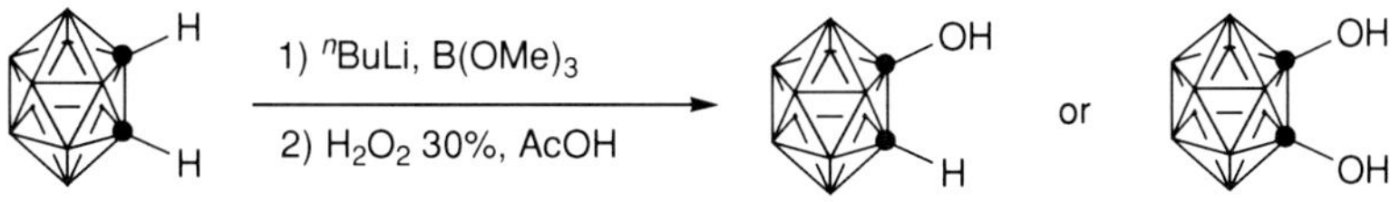

Scheme 1.11 Synthesis of C-hydroxylcarboranes

1.1.3 Reaction of Carboranylcopper

Another method was developed for the synthesis of 1,12-diethynylcarboranes or 1,12-diethenylcarboranes via carboranylcopper compounds. The authors also studied stereochemical aspects of Br$_2$, HCl, and HI addition to 1-ethynylcarboranes (Scheme 1.12) [33–36].

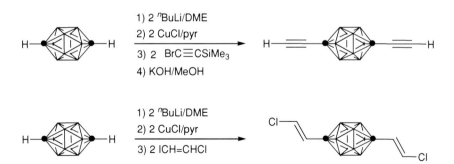

Scheme 1.12 Synthesis of diethynylcarboranes and diethenylcarboranes via carboranylcopper compounds

Carboranylcopper compounds can also react with arylhalides to give arylcarborane derivatives (Scheme 1.13) [37, 38].

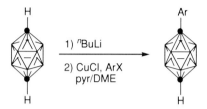

Scheme 1.13 Synthesis of C-arylcarboranes via carboranylcopper compounds

Reaction of dilithio-o-carborane with CuCl in toluene afforded a single product, 1,1′:2,2′-[Cu(toluene)]$_2$(C$_2$B$_{10}$H$_{10}$)$_2$, which gave 1,1′-bis(o-carborane) after hydrolysis. This serves as the most efficient method for the preparation of 1,1′-bis(o-carborane) (Scheme 1.14) [39].

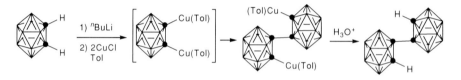

Scheme 1.14 Synthesis of 1,1′-bis(o-carborane)

The m-carborane was deprotonated with ^{n}BuLi and then treated with CuBr and LiBr followed by CS$_2$. Addition of MeI gave the dithioester, which was reduced by BH$_3$·SMe$_2$ to afford the thiol (Scheme 1.15) [40].

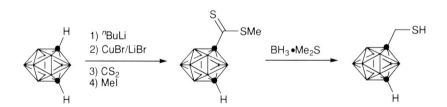

Scheme 1.15 Reaction of carboranylcopper compounds with CS$_2$

1.1 Synthesis of C-Substituted Carboranes

1.1.4 Other Methods

Phase transfer catalysis conditions appear to be more convenient in the preparation of some carborane derivatives. Kabachii et al. developed a method for the synthesis of 1,7-dichloro-*m*-carborane by chlorination of *m*-carborane with CCl_4 using phase transfer catalysts (Scheme 1.16) [41].

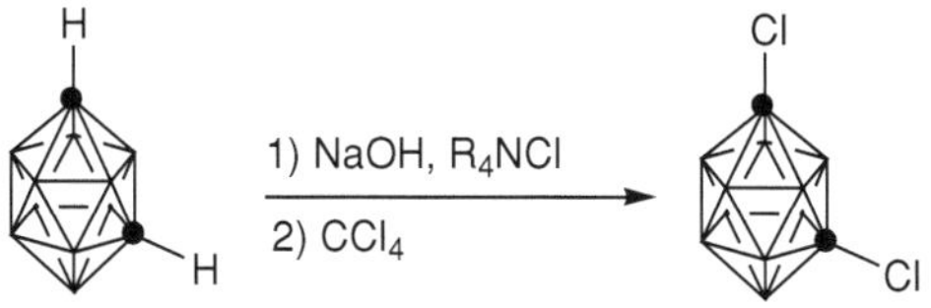

Scheme 1.16 Synthesis of C-chlorocarborane

Zakharkin synthesized 1-alkyl- and 1,2-di-alkyl-*o*-carboranes by alkylating *o*-carborane, and 1-methyl- or 1-phenyl-*o*-carborane with alkyl halides in alkali-THF system using dibenzo-18-crown-6-ether as transferring agent (Scheme 1.17) [42].

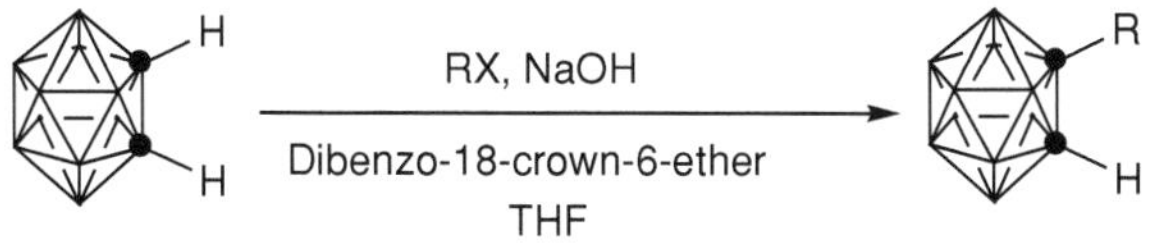

Scheme 1.17 Carborane alkylation in alkali-THF-(dibenzo-18-crown-6-ether) system

Yamamoto found that the addition of *o*-carborane to aldehydes proceeded very smoothly in the presence of aqueous tetrabutylammonium fluoride (TBAF), giving the corresponding carbinols in high yields. Furthermore, the TBAF-mediated reaction was applied to the [3+2] annulation between *o*-carborane (dianionic C_2 synthons) and α,β-unsaturated aldehydes and ketones (dicationic C_3 synthons) to give the corresponding five-membered carbocycles in good-to-high yields (Scheme 1.18) [43].

Scheme 1.18 TBAF promoted additions of *o*-carborane to aldehydes and enones

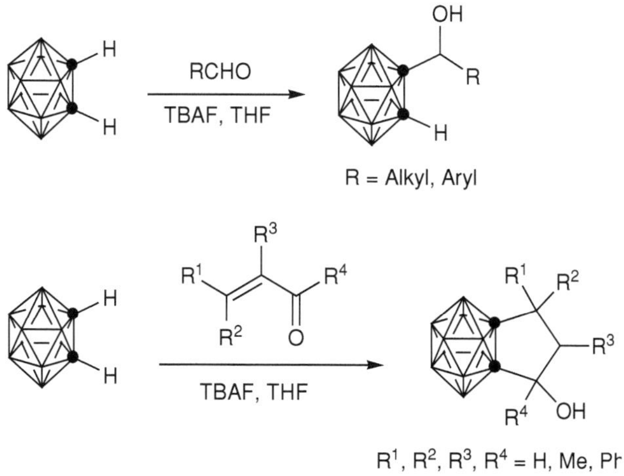

1.2 Synthesis of B-Substituted Carboranes

The chemistry of boron-substituted carboranes is not as developed as that of the carbon analogues due to the difficulty of introducing functional groups at the boron atom of the carborane cage. B-Halogenated carboranes, the first B-substituted species, appear to be inert to substitution reactions. B-Carboranyl compounds can be viewed as analogues of organic compounds because the B-carboranyl group acts as either an alkyl or aryl group in most transformations [16].

1.2.1 Reaction of Decaborane with Acetylenes

The first compound with a C–B(carborane) bond was obtained by the reaction of acetylene with a mixture of 1- and 2-ethyldecaboranes in acetonitrile, giving a mixture of 8- and 9-ethyl-*o*-carboranes [5].

1.2.2 Electrophilic Substitution of Carboranes

Another route to alkylcarboranes involves the electrophilic alkylation of carboranes with alkyl halide [44] or vinyltrichlorosilane [45] in the presence of AlCl$_3$ (Scheme 1.19).

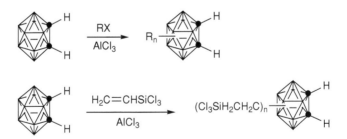

Scheme 1.19 Synthesis of B-alkylcarboranes

Direct electrophilic halogenation [46–50], alkylation [44, 45, 51], and metalation [52, 53] can take place at the boron atom (Scheme 1.20). These reactions are typical for aromatic compounds, and for this reason the carborane molecule has been termed as a "pseudoaromatic" system [54].

1.2 Synthesis of B-Substituted Carboranes

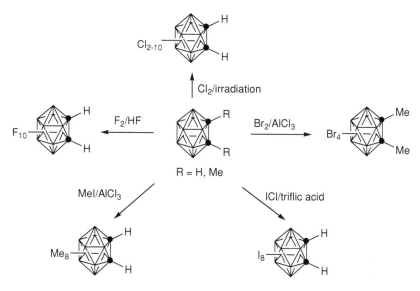

Scheme 1.20 Electrophilic substitution of carboranes

Theoretical calculations on carboranes show that electron density increases in the order 1 (2) < 3 (6) < 4 (5,7,11) < 8 (10) < 9 (12) for *o*-carborane, 1 (7) < 2 (3) < 5 (12) < 4 (6,8,11) < 9 (10) for *m*-carborane, and 1 (12) < 2 (3–11) for *p*-carborane (Scheme 1.2, positions listed in parentheses are chemically equivalent to those in front of the parentheses) [55–59]. Experimental results are in general agreement with theoretical calculations of the charge distribution. Electrophilic substitution usually occurs first at the 9,12 and then at the 8,10 positions of the *o*-carborane cage. The carbon atoms and the adjacent boron atoms do not appear susceptible to electrophilic substitution.

1.2.3 Reaction of Dicarbollide Ion ($C_2B_9H_{11}^{2-}$) with Dihaloborane

One of the most important reactions in carborane chemistry was reported by Wiesboeck and Hawthorne in 1964 [60, 61]. They showed that *o*-carborane could be degraded using alcoholic alkali removing one boron atom and forming the dicarbollide anion, $C_2B_9H_{11}^{2-}$. Starting from this anion, a number of 3-substituted *o*-caboranes were synthesized by the boron insertion reaction (Scheme 1.21) [62–67].

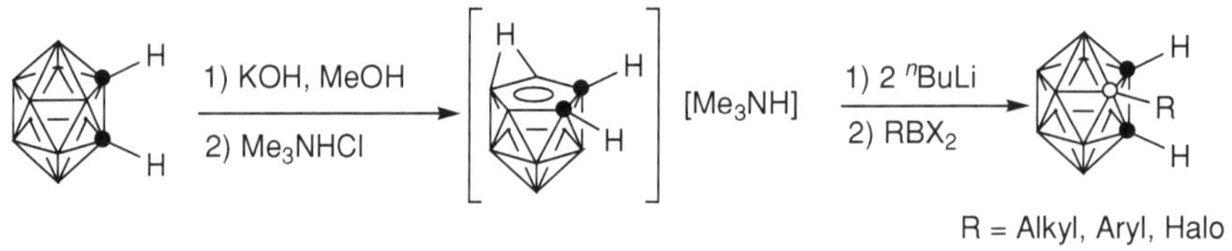

The image shows a chemical reaction scheme (Scheme 1.21).

Scheme 1.21 Synthesis of 3-substituted o-caboranes from $C_2B_9H_{11}^{2-}$

1.2.4 Transition Metal-Catalyzed Coupling Reaction

An organic group was introduced at the boron atom of the carborane cage through reaction of iodocarboranes with organomagnesium compounds in the presence of Ni or Pd complexes (Scheme 1.22) [68–72].

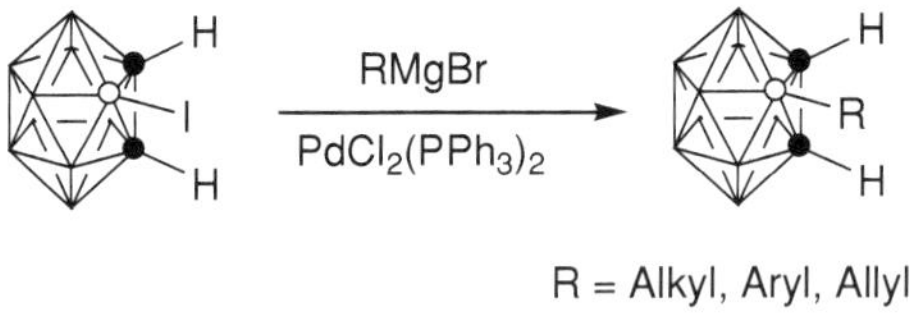

Scheme 1.22 Palladium-catalyzed reaction of 3-iodo-o-carborane with Grinard reagents

Methodology leading to a new class of rodlike p-carborane derivatives is described, involving the palladium-catalyzed coupling of B-iodinated p-carboranes with terminal alkynes (Scheme 1.23) [73]. The products of these reactions contain an alkyne substituent at a boron vertex of the p-carborane cage.

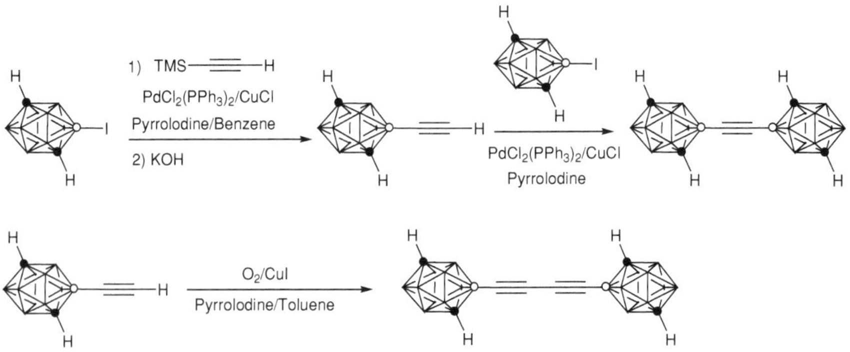

Scheme 1.23 Palladium-catalyzed coupling of B-iodo-p-carboranes with terminal alkynes

1.2 Synthesis of B-Substituted Carboranes

p-Carborane can be vinylated on the 2-*B*-atom in high yields using the Heck reaction (Scheme 1.24) [74]. Thus, the reaction between 2-iodo-*p*-carborane and various styrenes resulted in the production of the corresponding trans-β-(2-B-*p*-carboranyl)-styrene in DMF solution when reacted in the presence of silver phosphate and the palladacycle Herrmann's catalyst.

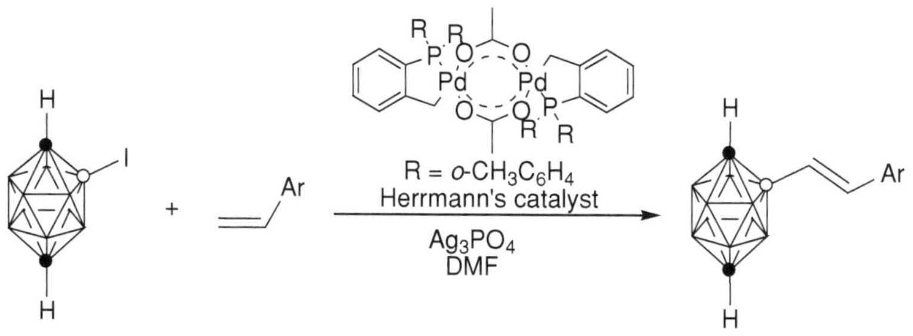

Scheme 1.24 Palladium-catalyzed coupling of B-iodo-*p*-carboranes with styrenes

The syntheses of 9-acetyl-*o*-carborane and 9-cyano-*o*-carborane are outlined in Scheme 1.25 [75]. Functionalization of *o*-carborane at the 9-position is readily achieved by iodination followed by reaction with the appropriate Grignard reagent. 9-Ethynyl-*o*-carborane and 9-ethyl-*o*-carborane were obtained by this route from 9-iodo-*o*-carborane. 9-Ethynyl-*o*-carborane is hydrated quantitatively in aqueous methanolic solution under catalysis by HgO and BF$_3$ with formation of 9-acetyl-*o*carborane. An acid obtained from oxidation of 9-ethyl-*o*-carborane with chromic anhydride, is allowed to react with thionyl chloride to give the corresponding acid chloride, which is converted into nitrile by reaction with sulfonylamide.

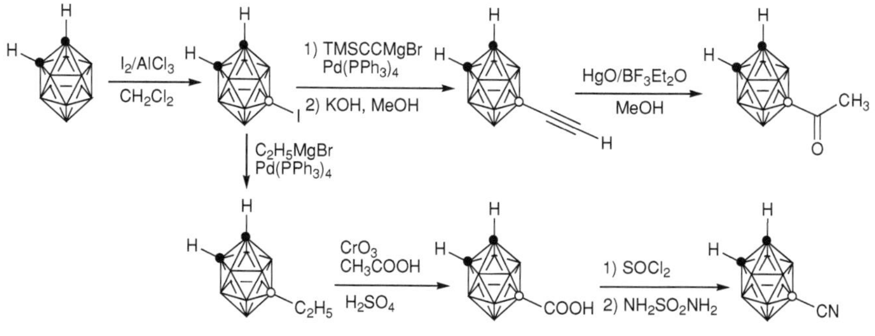

Scheme 1.25 Synthesis of 9-acetyl-*o*-carborane and 9-cyano-*o*-carborane

12   1   Introduction

1.2.5 Carbene Reaction

Jones et al. reported the reaction of carbomethoxycarbene with the B–H bonds of
o-carborane can form the products of formal B–H insertion, and the C–H bonds
were ignored by the carbene (Scheme 1.26) [76].

    An intramolecular version of this reaction can produce a series of carbon-
to-boron-bridged *o*-carboranes. The conversion of ketone to bridged benzo-
o-carborane is presented in Scheme 1.26 [77–79].

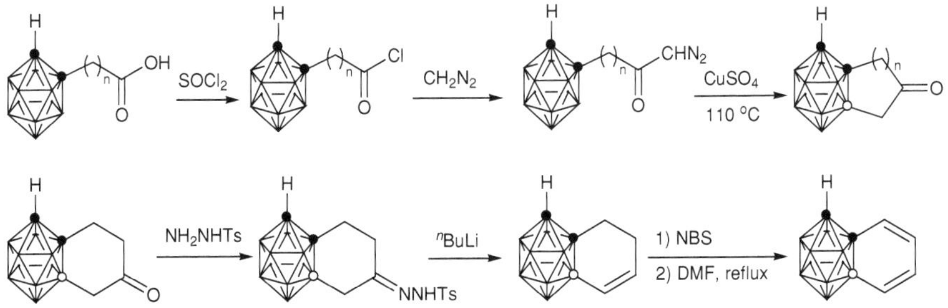

Scheme 1.26   B–H insertion of *o*-carborane with carbomethoxycarbene

1.2.6 Other Methods

Simple pyrolysis of *o*-carborane in the presence of dialkyl acetylenedicarboxylates
and trialkyl methane-tricarboxylates in sealed tubes at 275 °C produces 9-alkyl-
o-carboranes in reasonable yield (Scheme 1.27) [80].

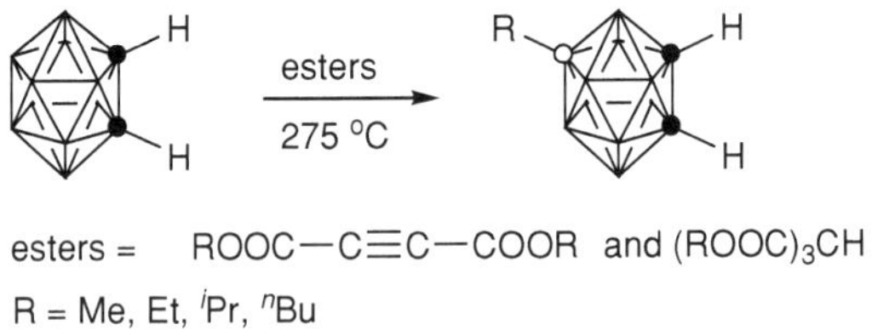

esters =   ROOC—C≡C—COOR   and (ROOC)$_3$CH

R = Me, Et, ^{i}Pr, ^{n}Bu

Scheme 1.27   Pyrolysis of *o*-carborane with dialkyl acetylenedicarboxylates and trialkyl
methane-tricarboxylates

    The first B-aminocarborane was obtained by Zakharkin and Kalinin in 1967
[81]. They showed that the dicarbadodecaborate anion, formed by the addition of
two electrons to the carborane nucleus, reacts with liquid ammonia at low

1.2 Synthesis of B-Substituted Carboranes 13

temperature and be oxidized with $KMnO_4$ or $CuCl_2$ to give 3-amino-o-carborane (Scheme 1.28). The 3-amino-o-carboranes show reactions typical of aliphatic and aromatic primary amines. They are readily arylated and acylated with formic acid or acetic anhydride.

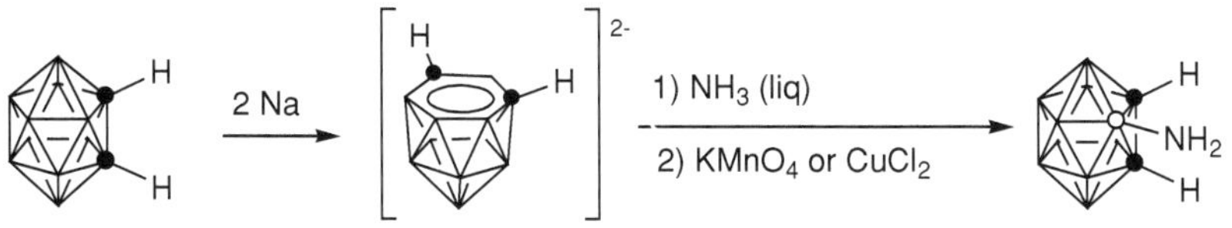

Scheme 1.28 Synthesis of 3-amino-o-carborane

The per-B-hydroxylated carboranes $closo$-1,12-H_2-1,12-$C_2B_{10}(OH)_{10}$, which may be considered to be derivatives of a new type of polyhedral subboric acid, can be prepared by the oxidation of the slightly water-soluble precursor $closo$-1,12-$(CH_2OH)_2$-1,12-$C_2B_{10}H_{10}$ with 30% hydrogen peroxide at the reflux temperature (Scheme 1.29), because $closo$-1,12-$C_2B_{10}H_{12}$ is water-insoluble and hence not available to the hydrogen peroxide reagent. During this reaction sequence, the diol is most likely oxidized to the corresponding dicarboxylic acid, which subsequently decarboxylates during B-hydroxylation [82, 83].

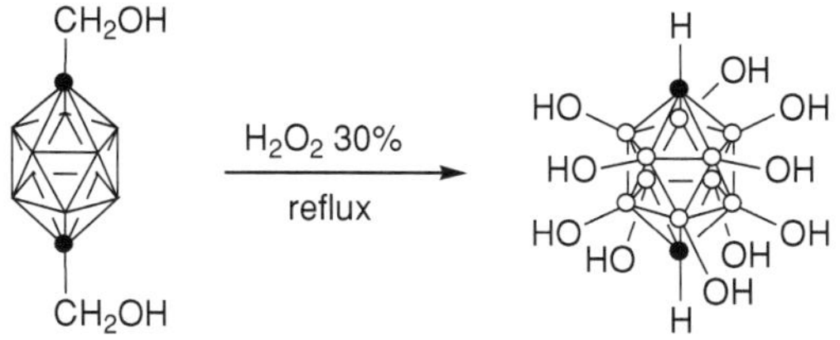

Scheme 1.29 Synthesis of per-B-hydroxylated carborane

1.3 1,2-o-Carboryne

Icosahedral carboranes, $closo$-$C_2B_{10}H_{12}$, are aromatic molecules which resemble benzene in both thermodynamic stability and chemical reactions. For example, carborane and benzene survive heating to several hundred degrees and undergo aromatic substitution reactions with electrophilic reagents [44–53]. Another dramatic aspect of benzene chemistry is the generation of benzyne which found many applications in organic synthesis, mechanistic studies, and synthesis of functional materials since its first report in 1950s [84–89]. Similar to benzene, o-carborane can also form this kind of dehydro-species, 1,2-o-carboryne (1,2-dehydro-o-carborane).

Jones and co-workers discovered a way to generate 1,2-*o*-carboryne by treatment of 1,2-dilithio-*o*-carborane with one equiv of bromine (Scheme 1.30) [90]. 1-Bromo-2-lithio-*o*-carborane is stable below 0 °C, however, upon heating in the presence of unsaturated molecules addition products are formed [91].

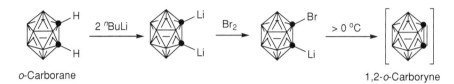

Scheme 1.30 Preparation of 1,2-*o*-carboryne intermediate from 1-Br-2-Li-1,2-C$_2$B$_{10}$H$_{10}$ precursor

When using diene as a trapping reagent, products of the [2+4] cycloaddition type, the [2+2] cycloaddition type, and ene reaction type are obtained with a very similar ratio to that of the reaction between benzyne and diene [91]. The mechanistic studies on these addition reactions show that both the [2+4]

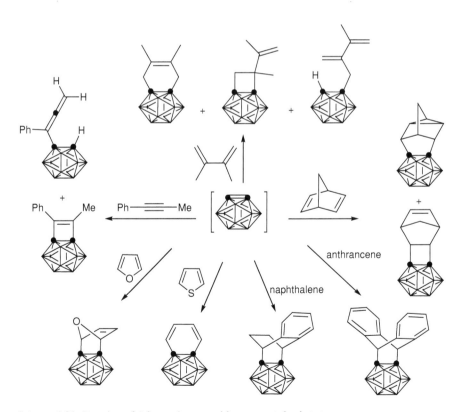

Scheme 1.31 Reaction of 1,2-*o*-carboryne with unsaturated substrates

1.3 1,2-o-Carboryne

cycloaddition and ene reaction are likely to be concerted, whereas the [2+2] cycloaddition might be stepwise. These are also similar to those of benzyne. Subsequently, the authors studied the reactions of 1,2-o-carboryne generated in situ with other dienes, alkynes, and alkenes, such as furans, thiophenes, anthracene, naphthalene, benzene, cyclohexene, norbornadiene, hexadiene and so on (Scheme 1.31) [90–96].

The benzene-1,2-o-carboryne adduct has been used as 1,2-o-carboryne precursor [96]. Under heating at 230–260 °C, the 1,2-o-carboryne moiety transfers from this adduct has been achieved in the presence of acceptors (Scheme 1.32). The naphthalene and anthracene adducts are thermally stable and cannot give similar result [96].

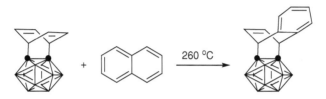

Scheme 1.32 Transfer of 1,2-o-carboryne moiety

Recently, a more efficient method has been developed for the generation of 1,2-o-carboryne under mild reaction conditions. Phenyl[o-(trimethylsilyl)carboranyl] iodonium acetate, prepared by the reaction of [o-(trimethylsilyl)carboranyl]lithium with IPh(OAc)$_2$, is such a kind of precursor (Scheme 1.33) [97, 98].

Scheme 1.33 Preparation of 1,2-o-carboryne intermediate from 1-TMS-2-IPh(OAc)-1,2-C$_2$B$_{10}$H$_{10}$ precursor

The reaction of this salt with anthracene in the presence of desilylating reagents such as CsF or KF/18-crown-6 gives 1,2-o-carboryne adduct in much higher yields. Other dienes such as naphthalene, 2,5-dimethylfuran and thiophene also work well as trapping reagents with improved yields compared with Jones' results. In a similar fashion, this precursor also functions well for the [2+2] addition reaction of 1,2-o-carboryne and strained cycloalkynes (Scheme 1.34) [98]. It should be mentioned that the cyclization of the in situ generated 1,2-o-carboryne with some alkynes in the presence of Ni(PEt$_3$)$_4$, Pd(PPh$_3$)$_4$ and Pt(PPh$_3$)$_4$ failed [98].

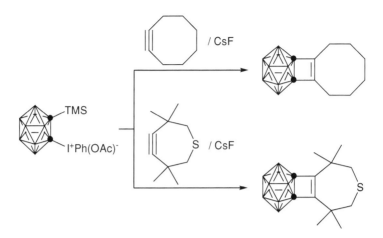

Scheme 1.34 Reaction of 1,2-*o*-carboryne with cycloalkynes

These experimental achievements spurred the theoretical study on this novel species. Although the experimental study only involves 1,2-C$_2$B$_{10}$H$_{10}$, the calculations encompass both 1,2-*o*-carboryne and 1,2-dehydro-*o*-silaboranes (*o*-C$_2$B$_n$H$_n$ and *o*-Si$_2$B$_n$H$_n$, n = 4, 5, 8, and 10) [99]. The study shows that the dehydrogeno formation of 1,2-C$_2$B$_{10}$H$_{10}$ is energetically comparable to that of benzyne with ca. 99 kcal/mol, whereas the dehydrogeno formation of 2,3-C$_2$B$_5$H$_5$ is estimated to be even less endothermic than that of 1,2-C$_2$B$_{10}$H$_{10}$ by more than 21 kcal/mol. The bond lengths of these dehydrogeno species are also calculated. For 1,2-C$_2$B$_{10}$H$_{10}$, the carbon–carbon bond length is 1.356 Å, which is shorter than that of 1.625 Å in 1,2-C$_2$B$_{10}$H$_{12}$, indicating the multiple bond character. This bond distance is still significantly longer than that observed in benzyne (1.245 Å). For 2,3-C$_2$B$_5$H$_5$, the bond distance of 1.305 Å is more comparable to that in benzyne. The calculation on the frontier molecular orbitals of the [4+2] cycloaddition between dienes and these 1,2-*o*-carboryne intermediate shows that the E(HOMO$_{\text{diene}}$-LUMO$_{\text{ene}}$) is lower than that for ethylene and benzyne. Since Diels–Alder reactions of ethylene, benzyne, and 1,2-C$_2$B$_{10}$H$_{10}$ with butadiene are known, other dehydrogeno carboranes are also expected to have similar reactivity [99].

1.4 Transition Metal-1,2-*o*-Carboryne Complexes

Transition metals were found capable of forming σ-bonds with the carbon atoms of the carborane cage. Derivatives with M–C(carborane) bonds are known for the following transition metals: Cu, Au, Ti, Zr, Mn, Re, Fe, Co, Rh, Ir, Ni, Pd, and Pt [16]. The organometallic derivatives of carboranes were mainly obtained from the reaction of lithiocarborane with compounds bearing metal-halogen bond (Scheme 1.35).

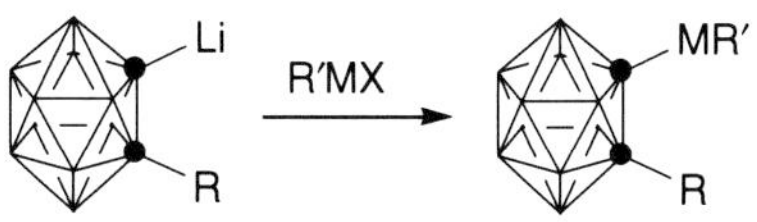

1.4 Transition Metal-1,2-o-Carboryne Complexes

Scheme 1.35 Synthesis of metal-carboranyl compounds

1.4.1 Synthesis of Transition Metal-1,2-o-Carboryne Complexes

The first example of transition metal-1,2-o-carboryne complexes, $(\eta^2$-$C_2B_{10}H_{10})$ Ni(PPh$_3$)$_2$, was reported in 1973 [100]. Treatment of 1,2-dilithio-o-caborane with MiCl$_2$(PPh$_3$)$_2$ (M=Ni, Pd, Pt) gives unique molecules (Ph$_3$P)$_2$M(η^2-$C_2B_{10}H_{10})$ (Scheme 1.36) [100, 101]. The structure of nickel complex was characterized by X-ray analysis. It contains a three-membered ring formed through two Ni–C(cage) bonds and the coordination plane about the nickel atom is essentially planar [100].

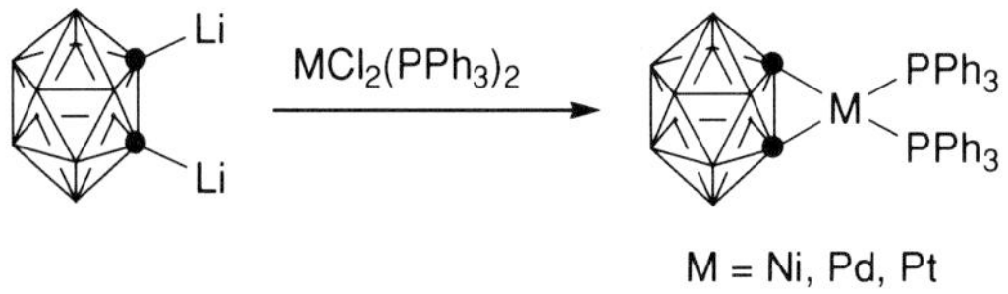

Scheme 1.36 Synthesis of group 10 metal-1,2-o-carboryne complexes

Ol'dekop et al. developed a decarboxylation procedure for the preparation of Ni-1,2-o-carboryne complexes stabilized by a bipyridyl ligand (Scheme 1.37) [102].

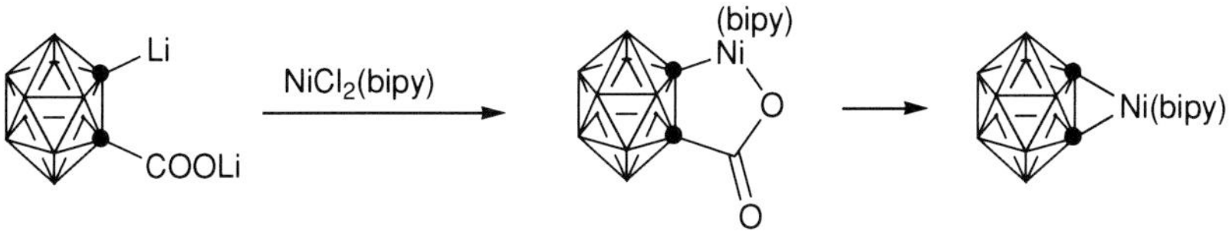

Scheme 1.37 Synthesis of Ni-1,2-o-carboryne complex by decarboxylation

Compounds with a Co–C(carborane) σ-bond were obtained by the interaction of lithiocarboranes with bipyridyl complexes of CoCl$_2$ (Scheme 1.38) [103].

Scheme 1.38 Synthesis of Co-1-o-carboranyl and Co-1,2-o-carboryne complexes

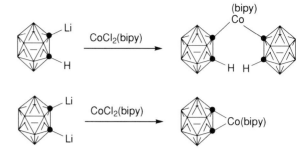

The first example of early transition metal-1,2-o-carboryne complex [{η^5:σ-Me$_2$C(C$_9$H$_6$)(C$_2$B$_{10}$H$_{10}$)}ZrCl(η^3-C$_2$B$_{10}$H$_{10}$)][Li(THF)$_4$] was prepared in 2003 from the reaction of in situ generated [η^5:σ-Me$_2$C(C$_9$H$_6$)(C$_2$B$_{10}$H$_{10}$)]ZrCl$_2$ with one equivalent of Li$_2$C$_2$B$_{10}$H$_{10}$ in THF in 60% yield (Scheme 1.39) [104]. Many attempts to remove the chloro ligand for the preparation of a neutral complex are not successful. The anionic nature of [{η^5:σ-Me$_2$C(C$_9$H$_{10}$)(C$_2$B$_{10}$H$_{10}$)}Zr(η^3-C$_2$B$_{10}$H$_{10}$)]$^-$ does not show any activity toward unsaturated molecules.

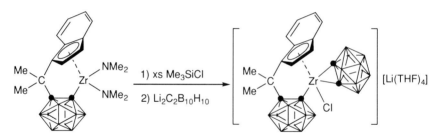

Scheme 1.39 Synthesis of [{η^5:σ-Me$_2$C(C$_9$H$_6$)(C$_2$B$_{10}$H$_{10}$)}ZrCl(η^3-C$_2$B$_{10}$H$_{10}$)][Li(THF)$_4$]

Single-crystal X-ray analyses show that the Zr atom is directly bonded to the two adjacent cage carbon atoms which do not have terminal hydrogen atoms. In addition, the metal center also interacts with the cage through an "agostic-like" Zr–H–B bond. Thus, the description Zr-η^3-(o-C$_2$B$_{10}$H$_{10}$) can be used to exemplify this novel bonding mode. With such a bonding description, the dianionic [η^3-(o-C$_2$B$_{10}$H$_{10}$)]$^{2-}$ ligand formally donates three pairs of electrons to the metal center and is isolobal with Cp$^-$. Therefore, one can conveniently correlate this zirconium complex anion with complexes having a general formula of d^0 Cp$_2$MX$_2$. Alternatively, one can describe the bonding interaction between the metal center and the two carbon atoms of the η^3-(o-C$_2$B$_{10}$H$_{10}$) ligand in terms of the metal–1,2-o-carboryne form. DFT calculations suggest that the bonding interactions between the Zr atom and 1,2-o-carboryne are best described as a resonance hybrid of both the Zr–C σ and Zr–C π bonding forms, similar to that observed in Cp$_2$Zr(η^2-benzyne) (Chart 1.1).

1.4 Transition Metal-1,2-o-Carboryne Complexes

Chart 1.1 Bonding interactions in Zr-benzyne and Zr-1,2-o-carboryne complexes

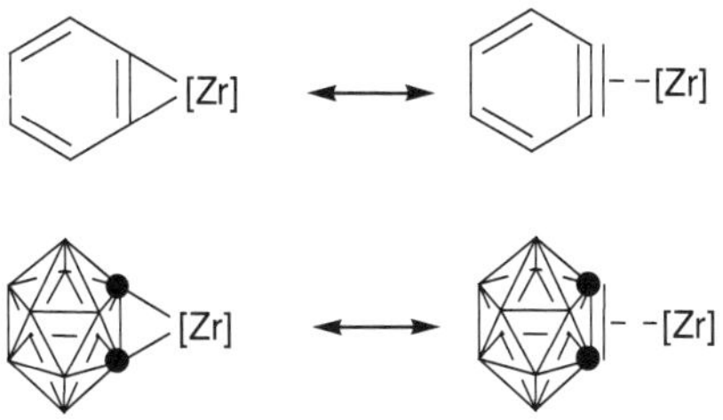

The salt metathesis between organozirconium dichloride and $Li_2C_2B_{10}H_{10}$ gave a class of zirconium-carboryne complexes, including $(\eta^2\text{-}C_2B_{10}H_{10})ZrCl_2(THF)_3$ (Scheme 1.40). Both the electronic and steric factors of the ligands have significant effects on the formation of the resultant metal complexes [105].

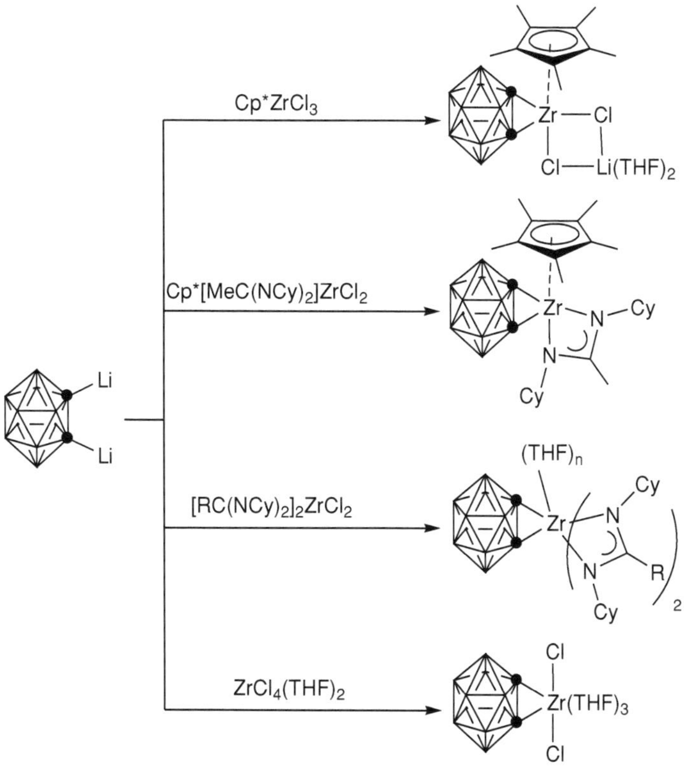

Scheme 1.40 Synthesis of Zr-1,2-o-carboryne complexes

Treatment of $(\eta^2\text{-}C_2B_{10}H_{10})ZrCl_2(THF)_3$ with 2 equiv of amidinatolithium, guanidinatolithium, or ^tBuOK afforded the complex $[\eta^2\text{-}CyNC(CH_3)NCy]_2Zr$ $(\eta^2\text{-}C_2B_{10}H_{10})$, $[\eta^2\text{-}{}^nPr_2NC(NPr^i)_2]_2Zr(\eta^2\text{-}C_2B_{10}H_{10})$, or $[(\eta^2\text{-}C_2B_{10}H_{10})_2Zr(O^tBu)$

(THF)][Zr(OBu^t)$_3$(THF)$_3$]. The unexpected product [σ:σ:σ-{^{t}BuC(O)=CHC(^{t}Bu)(O)-C$_2$B$_{10}$H$_{10}$}]Zr(η^2-^{t}BuCOCHCOBu^t)(THF)$_2$ was isolated from the reaction of (η^2-C$_2$B$_{10}$H$_{10}$)ZrCl$_2$(THF)$_3$ with (^{t}BuCOCHCO^{t}Bu)Na (Scheme 1.41) [105].

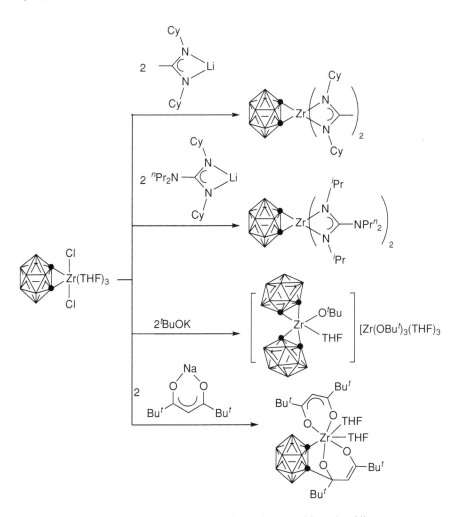

Scheme 1.41 Reaction of dichlorozirconium-1,2-*o*-carboryne with nucleophiles

1.4.2 Reactivity of Zr-1,2-o-Carboryne Complexes

Attempts to synthesize Cp$_2$Zr(η^2-C$_2$B$_{10}$H$_{10}$)(L), an analogue of Cp$_2$Zr(η^2-C$_6$H$_4$)(L), via treatment of Cp$_2$ZrCl$_2$ with one equivalent of Li$_2$C$_2$B$_{10}$H$_{10}$ fail. This reaction gives, instead, the ate-complex Cp$_2$Zr(μ-Cl)(μ-C$_2$B$_{10}$H$_{10}$)Li(OEt$_2$)$_2$ (**I-1**) in 70% isolated yield (Scheme 1.42) [106]. Complex **I-1** can be viewed as a precursor of zirconocene-carboryne Cp$_2$Zr(η^2-C$_2$B$_{10}$H$_{10}$).

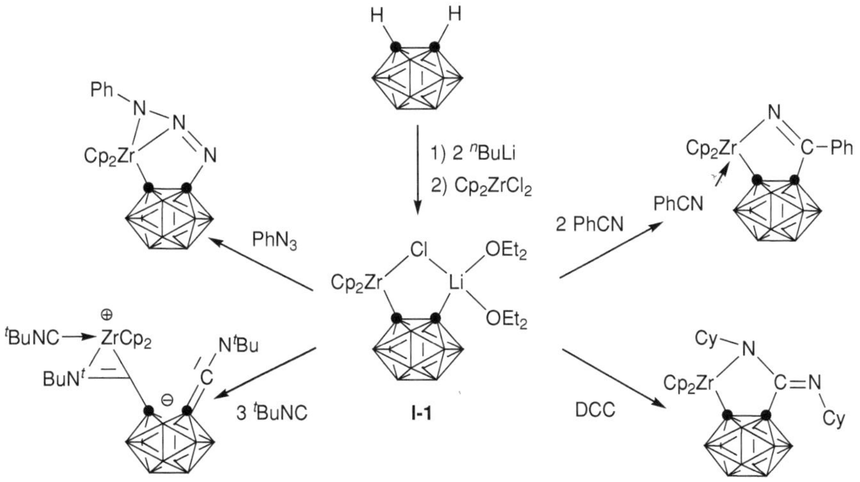

1.4 Transition Metal-1,2-o-Carboryne Complexes

Scheme 1.42 Reactivity of **I-1**

Treatment of **I-1** with PhCN, CyN=C=NCy, PhN$_3$, and ^{t}BuNC affords the insertion products Cp$_2$Zr[σ:σ-N=C(Ph)(C$_2$B$_{10}$H$_{10}$)](PhCN), Cp$_2$Zr[σ:σ-CyNC(=NCy)-(C$_2$B$_{10}$H$_{10}$)], Cp$_2$Zr[η^2:σ-(PhNN=N)(C$_2$B$_{10}$H$_{10}$)], and Cp$_2$Zr[η^2-^{t}BuNC(C$_2$B$_{10}$H$_{10}$)=CN^{t}Bu](CN^{t}Bu), respectively, in moderate to high yields (Scheme 1.42) [106].

Scheme 1.43 shows the proposed reaction mechanism. Dissociation of LiCl from **I-1** gives the key intermediate Cp$_2$Zr(η^2-C$_2$B$_{10}$H$_{10}$). Coordination of PhCN and subsequent insertion generate the five-membered metallacycle. The coordination sphere of the Zr atom is then completed by binding to another equivalent of PhCN molecule. No further insertion proceeds because of the steric reasons. For ^{t}BuNC, the first insertion of ^{t}BuNC into the Zr–C(cage) bond gives a four-membered metallacycle, followed by the further insertion of the second molecule of ^{t}BuNC to afford a five-membered metallacycle. The coordination of the imine nitrogen and the cleavage of one Zr–C(imine) bond lead to the production of the final product. Back-donation of the carboanion to the cage carbon can lead to the formation of *exo* C(cage)=C double bond and the subsequent cleavage of the cage C–C bond [106].

Complex **I-1** can also react with various kinds of alkynes, leading to the formation of metallacyclopentenes. An equimolar reaction of **I-1** with RC≡CR in refluxing toluene gives 1,2-[Cp$_2$ZrC(R)=C(R)]-1,2-C$_2$B$_{10}$H$_{10}$ (**I-2**) in very high isolated yield (Scheme 1.44) [107]. An alkyne-coordinated complex is suggested to be the intermediate. The polarity of alkynes determines the regioselectivity of the insertion products. Like the reaction with polar unsaturated molecules, no further insertion products are detected even after prolonged heating in the presence of an excess amount of alkynes [107].

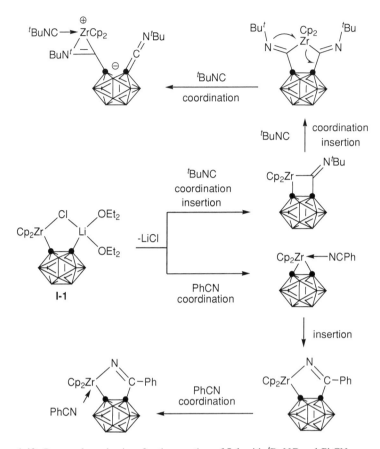

Scheme 1.43 Proposed mechanism for the reaction of **I-1** with ^{t}BuNC and PhCN

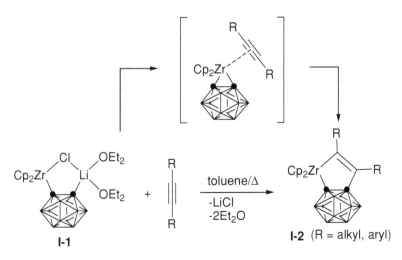

Scheme 1.44 Reaction of **I-1** with alkynes

1.4 Transition Metal-1,2-*o*-Carboryne Complexes 23

Complex **I-2** are very useful starting materials for the preparation of functionalized carboranes (Scheme 1.45). Hydrolysis under acidic media affords alkenylcarborane 1-[HC(Et)=C(Et)]-1,2-$C_2B_{10}H_{11}$. Interaction of **I-2** with I_2 in the presence of CuCl generates a monosubstituted carborane 1-[CI(Et)=C(Et)]-1,2-$C_2B_{10}H_{11}$ in 71% isolated yield. Disubstituted species 1-I-2-[CI(Et)=C(Et)]-1,2-$C_2B_{10}H_{10}$ is not observed. This result is very different from that of its analogue zirconacyclopentadienes $Cp_2Zr[C(R)=C(R)-C(R)=C(R)]$, in which the diiodo species is the major product in the presence of CuCl. Therefore, it is rational to suggest that, after transmetalation to Cu(I), only the Cu–C(vinyl) bond is reactive toward I_2 whereas the Cu–C_{cage} bond is inert probably because of steric reasons. Reaction of **I-2** with *o*-diiodobenzene in the presence of CuCl produces naphthalocarborane 1,2-[*o*-$C_6H_4C(Et)=C(Et)$]-1,2-$C_2B_{10}H_{10}$ in 81% isolated yield. Treatment of **I-2** with $CuCl_2$ in toluene at 80 °C gives the C–C coupling product 1,2-[C(Et)=C(Et)]-1,2-$C_2B_{10}H_{10}$. 2,6-$(CH_3)_2C_6H_3NC$ can readily insert into the Zr–C_{vinyl} bond to form an insertion product 1,2-[(2′,6′-$Me_2C_6H_3N=$)CC(Et)=C(Et)]-1,2-$C_2B_{10}H_{10}$ in refluxing toluene in the absence of CuCl [108].

Scheme 1.45 Reactivity of **I-2**

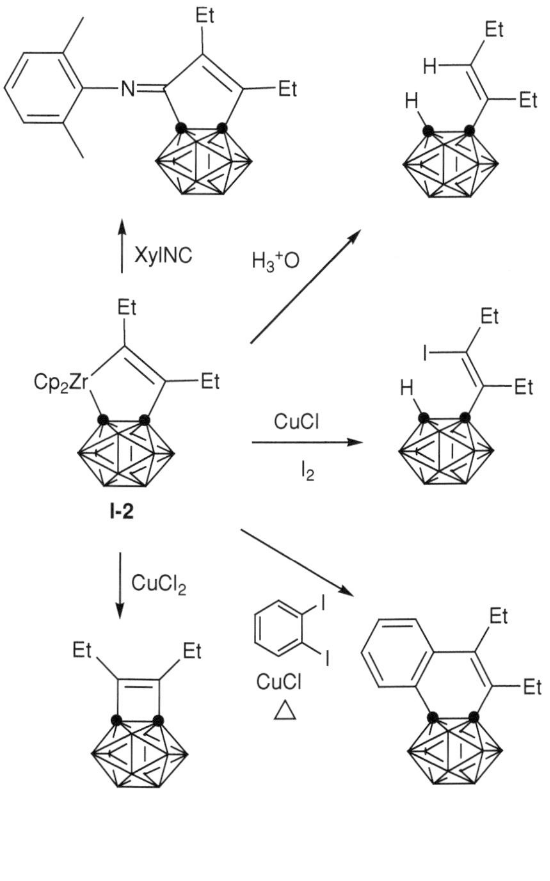

1.4.3 Reactivity of Ni-1,2-o-Carboryne Complexes

In view of the reactions of nickel-benzyne with alkynes to generate substituted naphthalenes and the analogy between nickel-benzyne and nickel-1,2-o-carboryne complex (Chart 1.2), the reactivity of $(\eta^2\text{-}C_2B_{10}H_{10})Ni(PPh_3)_2$ was examined.

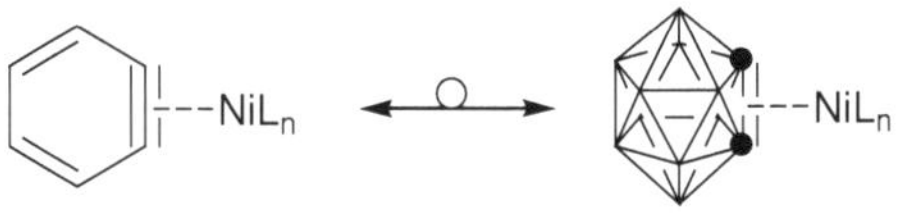

Chart 1.2 Isolobal analogue

Structural data of $(\eta^2\text{-}C_2B_{10}H_{10})Ni(PPh_3)_2$ show that the C_{cage}–C_{cage} bond distance in $(\eta^2\text{-}C_2B_{10}H_{10})Ni(PPh_3)_2$ is shorter than the corresponding value observed in Zr-1,2-o-carboryne complex [100], suggestive of the effects of electronic configuration of the metal center on the bonding interactions between the metal atom and carboryne unit. As a result, $(\eta^2\text{-}C_2B_{10}H_{10})Ni(PPh_3)_2$ does not react with any polar unsaturated molecules, but it reacts well with alkynes.

Treatment of $(\eta^2\text{-}C_2B_{10}H_{10})Ni(PPh_3)_2$ with internal alkynes gives highly substituted benzocarboranes 1,2-[C(R^1)=C(R^2)C(R^1)=C(R^2)]-1,2-C$_2$B$_{10}$H$_{10}$ via a [2+2+2] cycloaddition (Scheme 1.46) [109]. The formation of benzocarborane can

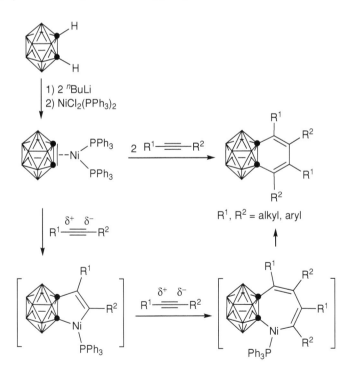

Scheme 1.46 Reaction of Ni-1,2-o-carboryne with alkynes

1.4 Transition Metal-1,2-o-Carboryne Complexes

be rationalized by the sequential insertion of alkynes into the Ni–C bond; followed by reductive elimination. The first insertion into the Ni–C(cage) bond gives a nickelacyclopentene intermediate. The exclusive formation of the head-to-tail products suggests that the insertion of the second equivalent of alkyne into the Ni–C(vinyl) bond is highly preferred over the Ni–C(cage) bond, leading to the regioselective products.

1.5 Our Objectives

In view of the rich chemistry displayed by the 1,2-o-carboryne and its transition metal complexes, the research objectives of this research are (1) synthesis of new nickel-1,2-o-carboryne complexes, (2) exploration of the reaction chemistry of Ni-1,2-o-carboryne, and (3) development of new transition metal-1,3-o-carboryne chemistry. In the following chapters of this thesis, we would like to describe the details of our efforts on these subjects.

A series of B-substituted nickel-1,2-o-carboryne complexes, $(\eta^2-1,2-C_2 B_{10}R_n^1H_{10-n})Ni(PR_3^2)_2$, were synthesized by salt elimination of phosphine ligated metal halide with dilithiocarboranes. Both the substituents on the carborane cage and the phosphine ligands have significant effects on the stability of these complexes.

The reactivity of $(\eta^2-C_2B_{10}H_{10})Ni(PPh_3)_2$ toward alkenes was studied and a novel nickel-mediated coupling reaction of carboryne with a variety of alkenes was developed, which gives alkenylcarboranes in moderate to very good isolated yields with excellent regio- and stereoselectivity. The intramolecular coordination of the heteroatom in alkenes can suppress β-H elimination reactions, leading to the isolation of the thermodynamically stable inserted intermediates, [{[2-CH$_2$ CH(o-C$_5$H$_4$N)-1,2-C$_2$B$_{10}$H$_{10}$]Ni}$_3(\mu_3$-Cl)][Li(DME)$_3$] and [2-CH$_2$CH(CO$_2$Me)-1, 2-C$_2$B$_{10}$H$_{10}$]Ni(PPh$_3$). These intermediates react readily with alkynes to give three-component [2+2+2] cycloaddition products. A novel nickel-mediated three-component assembling reaction of carboryne with alkenes and alkynes was then developed to give corresponding dihydro-1,2-benzo-o-carboranes. Accordingly, a new method for the synthesis of 1,2-dihydronaphthalenes from readily available starting materials also was developed, which involves nickel-catalyzed carboannulation of arynes, activated alkenes, and alkynes. The formation of these products can be rationalized by the sequential insertion of alkene and alkyne into the Ni–C bond.

A catalytic version of [2+2+2] cycloaddition reaction of carboryne with alkynes was achieved using 1-iodo-2-lithiocarborane as precursor and NiCl$_2$(PPh$_3$)$_2$ as catalyst. The mechanism involved oxidative addition of Ni into the cage C–I bond, elimination of LiI to form Ni-1,2-o-carboryne, and sequential alkyne insertion into the Ni–C$_{cage}$ bond and Ni–C$_{vinyl}$ bond, followed by reductive elimination, was proposed after the NMR reaction study and the structural confirmation of the key intermediate, nickelacyclopentene $[\{[2-C(^nBu)=C(o-C_5H_4N)-1,2-C_2B_{10}H_{10}]Ni\}_2 (\mu_2-Cl)][Li(THF)_4]$, from the reaction of n-butyl-2-pyridinylacetylene.

1,3-Dehydro-o-carborane was observed for the first time, which can be trapped by unsaturated molecules in the presence of a catalytic amount of transition metal. This leads to a discovery of a palladium/nickel-cocatalyzed [2+2+2] cycloaddition reaction of 1,3-o-carboryne with alkynes affording 1,3-benzo-o-carboranes. This work offers a new methodology for B-functionalization of carborane and demonstrates the relative reactivity of M–C over M–B bond in metal-1,3-o-carboryne complexes toward alkynes.

These methodologies provide exceptionally efficient routes from readily available starting materials to a wide variety of functionalized carboranes, which have potential use in medicinal and materials chemistry.

References

1. Hawthorne MF (1993) Angew Chem Int Ed Engl 32:950
2. Soloway AH, Tjarks W, Barnum JG (1998) Chem Rev 98:1515
3. Valliant JF, Guenther KJ, King AS, Morel P, Schaffer P, Sogbein OO, Stephenson KA (2002) Coord Chem Rev 232:173
4. Armstrong AF, Valliant JF (2007) Dalton Trans 38:4240–4251
5. Heying TL, Ager JW Jr, Clark SL, Mangold DJ, Goldstein HL, Hillman M, Polak RJ, Szymanski JW (1963) Inorg Chem 2:1089
6. Fein MM, Bobinski J, Mayes N, Schwartz N, Cohen MS (1963) Inorg Chem 2:1111
7. Zakharkin LI, Stanko VI, Brattsev VA, Chapovsky Yu A, Stuchkov Yu T (1963) Izu Akad Nauk SSSR, Ser Khim, 2069
8. Zakharkin LI, Stanko VI, Brattsev VA, Chapovsky Yu A, Okhlobystin Yu O (1963) Izu Akad Nauk SSSR, Ser Khim, 2069
9. Grimes RN (1970) Carboranes. Academic, New York
10. GMELIN Institute for Inorganic Chemistry of the Max-Planck-Society for the Advancement of Science (1988) Gmelin Handbook of Inorganic Chemistry, vol. 4, 8th edn. 3rd Suppl, Springer, Berlin
11. Bresadola S (1982) In: Grimes RN (ed) Metal interactions with boron clusters. Plenum Press, New York
12. Onak T (1995) Polyhedral Carbaboranes. In: Abel EW, Stone FGA, Wilkinson G (eds) Comprehensive organometallic chemistry II, vol 1. Pergamon, New York, p 217
13. Grimes RN (1995) Metallacarboranes. In: Abel EW, Stone FGA, Wilkinson G (eds) Comprehensive organometallic chemistry II, vol 1. Pergamon, New York, p 373
14. Davidson M, Hughes AK, Marder TB, Wade K (2000) Contemporary boron chemistry. Royal Society of Chemistry, Cambridge
15. Bubnov YN (2002) Boron chemistry at the beginning of the 21st century. Russian Academy of Sciences, Moscow
16. Bregadze VI (1992) Chem Rev 92:209
17. Štíbr B (1992) Chem Rev 92:225
18. Siebert W, Maier C-J, Maier A, Greiwe P, Bayer MJ, Hofmann M, Pritzkow H (2003) Pure Appl Chem 75:1277
19. Grimes RN (2003) Pure Appl Chem 75:1211
20. Hosmane NS, Maguire JA (2005) Organometallics 24:1356

References

21. Beall H (1972) Inorg Chem 11:637
22. Valliant JF, Schaffer P, Stephenson KA, Britten JF (2002) J Org Chem 67:383
23. Kusari U, Li Y, Bradley MG, Sneddon LG (2004) J Am Chem Soc 126:8662
24. Li Y, Carroll PJ, Sneddon LG (2008) Inorg Chem 47:9193
25. Grafstein D, Bobinski J, Dvorak J, Smith HF, Schwartz NN, Cohen MS, Fein MM (1963) Inorg Chem 2:1120
26. Heying TL, Ager JW, Clark SL, Alexander RP, Papetti S, Reid JA, Trotz SI (1963) Inorg Chem 2:1097
27. Gomez FA, Johnson SE, Hawthorne MF (1991) J Am Chem Soc 113:5915
28. L'Esperance RP, Li Z, Engen DV, Jones M Jr (1989) Inorg Chem 28:1823
29. Barnett-Thamattoor L, Wu JJ, Ho DM, Jones M Jr (1996) Tetrahedron Lett 37:7221
30. Kahl SB, Kasar RA (1996) J Am Chem Soc 118:1223
31. Dozzo P, Kasar RA, Kahl SB (2005) Inorg Chem 44:8053
32. Ohta K, Goto T, Yamazaki H, Pichierri F, Endo Y (2007) Inorg Chem 46:3966
33. Zakharkin LI, Kovredov AI, Ol'shevskaya VA (1986) Izu Akad Nauk SSSR, Ser Khim, 1388
34. Batsanov AS, Fox MA, Howard JAK, MacBride JAH, Wade K (2000) J Organomet Chem 610:20
35. Fox MA, Baines TE, Albesa-Jové D, Howard JAK, Low PJ (2006) J Organomet Chem 691:3889
36. Fox MA, Cameron AM, Low PJ, Paterson MAJ, Batsanov AS, Goeta AE, Rankin DWH, Robertsonb HE, Schirlinb JT (2006) Dalton Trans, 3544
37. Fujii S, Goto T, Ohta K, Hashimoto Y, Suzuki T, Ohta S, Endo Y (2005) J Med Chem 48:4654
38. Ghirotti M, Schwab PFH, Indelli MT, Chiorboli C, Scandola F (2006) Inorg Chem 45:4331
39. Ren S, Xie Z (2008) Organometallics 27:5167
40. Todd JA, Turner P, Ziolkowski EJ, Rendina LM (2005) Inorg Chem 44:6401
41. Kabachii Yu A, Valetakii PM, Vinogradova SV, Korshak VV (1985) Izu Akad Nauk SSSR, Ser Khim, 1932
42. Zakharkin LI, Ol'shevskaya VA, Kobak VV, Boiko NB (1988) Metalloorg Khim 1:364
43. Nakamura H, Aoyagi K, Yamamoto Y (1998) J Am Chem Soc 120:1167
44. Zakharkin LI, Pisareva IV, Bikkineev R (1977) Kh Izv Akad Nauk SSSR, Ser Khim, 641
45. Mironov VF, Grigos VI, Pechurina SYa, Zhigach AF (1972) Zh Obshch Khim 42:2583
46. Andrew JS, Zayas J, Jones M Jr (1985) Inorg Chem 24:3715
47. Zakharkin LI, Kalinin VN (1966) Izv Akad Nauk SSSR, Ser Khim, 575
48. zakharkin LI, Ol'Shevskaya VA, Poroahin Yu T, Balgurova EV (1987) Zh Obshch Khim 57:2012
49. Lebedev VN, Balagurova EV, Polyakov AV, Yanowky AI, Struchkov Yu T, Zakharkin LI (1990) J Organomet Chem 385:307
50. Teixidor F, Barberà G, Viñas C, Sillanpää R, Kivekäs R (2006) Inorg Chem 45:3496
51. Herzog A, Maderna A, Harakas GN, Knobler CB, Hawthorne MF (1999) Chem Eur J 5:1212
52. Bregadze VI, Kampel VTs, Godovikov NN (1976) J Organomet Chem 112:249
53. Bregadze VI, Usiatinsky AYa (1985) J Organomet Chem 292:75
54. King RB (2001) Chem Rev 101:1119
55. Potenza JA, Lipscomb WN, Vickers GD, Schroeder HA (1966) J Am Chem Soc 88:628
56. Hoffmann R, Lipscomb WN (1962) J Chem Phys 36:2179
57. Koetzle TF, Lipscomb WN (1970) Inorg Chem 9:2743
58. Beall H, Lipscomb WN (1967) Inorg Chem 6:874
59. Dixon DA, Kleier DA, Helgren TA, Hall JH, Lipscomb WN (1977) J Am Chem Soc 99:6226
60. Wiesboeck RA, Hawthorne MF (1964) J Am Chem Soc 86:1642
61. Barberà G, Vaca A, Teixidor F, Sillanpää R, Kivekäs R, Viñas C (2008) Inorg Chem 47:7309
62. Hawthorne MF, Wegner PA (1965) J Am Chem Soc 87:4302

63. Hawthorne MF, Wegner PA (1968) J Am Chem Soc 90:896
64. Potapova TV, Mikhailov BM (1967) Izu Akad Nauk SSSR, Ser Khim, 2367
65. Roscoe JS, Kongpricha S, Papetti S (1970) Inorg Chem 9:1561
66. Li J, Jones M Jr (1990) Inorg Chem 29:4162
67. Li J, Caparrelli DJ, Jones M Jr (1993) J Am Chem Soc 115:408
68. Zakharkin LI, Kovredov AI, Ol'shevskaya VA, Shaugumbekova ZhS (1980) Izv Akad Nauk SSSR, Ser Khim, 1691
69. Kovredov AI, Shaugumbekova ZhS, Petrovskii PV, Zakharkin LI (1989) Zh Obshch Khrm 59:607
70. Jiang W, Knobler CB, Curtis CE, Mortimer MD, Hawthorne MF (1995) Inorg Chem 34:3491
71. Clara Vinas C, Barberà G, Oliva JM, Teixidor F, Welch AJ, Rosair GM (2001) Inorg Chem 40:6555
72. Mukhin SN, Kabytaev KZ, Zhigareva GG, Glukhov IV, Starikova ZA, Bregadze VI, Beletskaya IP (2008) Organometallics 27:5937
73. Jiang W, Harwell DE, Mortimer MD, Knobler CB, Hawthorne MF (1996) Inorg Chem 35:4355
74. Eriksson L, Winberg KJ, Claro RT, Sjöberg S (2003) J Org Chem 68:3569
75. Crăciun L, Custelcean R (1999) Inorg Chem 38:4916
76. Zheng G, Jones M Jr (1983) J Am Chem Soc 105:6487
77. Wu S, Jones M Jr (1986) Inorg. Chem 25:4802
78. Wu S, Jones M Jr (1988) Inorg Chem 27:2005
79. Wu S, Jones M Jr (1989) J Am Chem Soc 111:5373
80. Zakharkin LI, Kalinin VN (1967) Izv Akad Nauk SSSR, Ser Khim, 2585
81. Albagli D, Zheng G-X, Jones M Jr (1986) Inorg Chem 25:129
82. Peymann F, Herzog A, Knobler CB, Hawthorne MF (1999) Angew Chem Int Ed 38:1061
83. Herzog A, Knobler CB, Hawthorne MF (2001) J Am Chem Soc 123:12791
84. Roberts JD, Simmons HE Jr, Carlsmith LA, Vaughan CW (1953) J Am Chem Soc 75:3290
85. Hart H (1994) In: Patai S (ed) Chemistry of triple-bonded functional groups, Supplement C2, Chap. 18. Wiley, Chichester
86. Hoffmann RW (1967) Dehydrobenzene and Cycloalkynes. Academic, New York
87. Gilchrist TL (1983) In: Patai S, Rappoport Z (eds) Chemistry of functional groups, Supplement C, Chap. 11. Wiley, Chichester
88. Buchwald SL, Nielsen RB (1988) Chem Rev 88:1047
89. Jones WM, Klosin J (1998) Adv Organomet Chem 42:147
90. Gingrich HL, Ghosh T, Huang Q, Jones M Jr (1990) J Am Chem Soc 112:4082
91. Ghosh T, Gingrich HL, Kam CK, Mobraaten ECM, Jones M Jr (1991) J Am Chem Soc 113:1313
92. Huang Q, Gingrich HL, Jones M Jr (1991) Inorg Chem 30:3254
93. Cunningham RT, Bian N, Jones M Jr (1994) Inorg Chem 33:4811
94. Ho DM, Cunningham RJ, Brewer JA, Bian N, Jones M Jr (1995) Inorg Chem 34:5274
95. Barnett-Thamattoor L, Zheng G, Ho DM, Jones M Jr, Jackson JE (1996) Inorg Chem 35:7311
96. Atkins JH, Ho DM, Jones M Jr (1996) Tetrahedron Lett 37:7217
97. Lee T, Jeon J, Song KH, Jung I, Baik C, Park K-M, Lee SS, Kang SO, Ko J (2004) Dalton Trans 933
98. Jeon J, Kitamura T, Yoo B-W, Kang SO, Ko J (2001) Chem Commun 20:2110
99. Kiran B, Anoop A, Jemmis ED (2002) J Am Chem Soc 124:4402
100. Sayler AA, Beall H, Sieckhaus JF (1973) J Am Chem Soc 95:5790
101. Zakharkin LI, Kovredov AI, (1975) IzvAkad Nauk SSSR, Ser Khim, 2619
102. Ol'dekop Yu A, Maier NA, Erdman AA, Prokopovich VP (1981) Dokl Akad Nauk SSSR 257:647
103. Ol'dekop Yu A, Maier NA, Erdman AA, Prokopovich VP (1982) Zh Obshch Khim 52:2256
104. Wang H, Li H-W, Huang X, Lin Z, Xie Z (2003) Angew Chem Int Ed 42:4347

References

105. Ren S, Deng L, Chan H-S, Xie Z (2009) Organometallics 28:5749
106. Deng L, Chan H-S, Xie Z (2005) J Am Chem Soc 127:13774
107. Ren S, Chan H-S, Xie Z (2009) Organometallics 28:4106
108. Ren S, Chan H-S, Xie Z (2009) J Am Chem Soc 131:3862
109. Deng L, Chan H-S, Xie Z (2006) J Am Chem Soc 128:7728

Chapter 2
Nickel-1,2-*o*-Carboryne Complexes

2.1 Introduction

1,2-*o*-Carboryne (1,2-dehydro-*o*-carborane), which was first reported as a reactive intermediate in 1990 [1], is very energetically comparable with its two-dimensional relative benzyne [2]. Reactivity studies also showed that they are quite similar in reactions with unsaturated molecules [3–9].

Like benzyne, carboryne can be trapped and stabilized by transition metals. The reaction of organozirconium dichloride with 1 equiv of $Li_2C_2B_{10}H_{10}$ or treatment of (η^2-$C_2B_{10}H_{10}$) $ZrCl_2(THF)_3$ with anionic ligands can give a class of zirconium-1,2-*o*-carboryne complexes [10]. Molecular orbital calculations suggested that the bonding interactions between Zr and 1,2-*o*-carboryne are best described as a resonance hybrid of both Zr–C σ and Zr–C π bonding forms which is similar to that observed in Zr-benzyne complex [11]. For late transition metals, salt metathesis is also a good method for the synthesis of metal-1,2-*o*-carboryne complexes by reaction of MCl_2 (M = Ni, Pd, Pt, Co) with $Li_2C_2B_{10}H_{10}$ [12–14]. A series of Ni-1,2-*o*-carboryne was recently prepared in our group by the reaction of phosphine ligated nickel halide. The C(cage)–C(cage) bonds in these complexes are much shorter than those in Zr-1,2-*o*-carboryne complexes [10]. The reactivity studies on (η^2-$C_2B_{10}H_{10}$)Ni(PPh$_3$)$_2$ show that the coordinated PPh$_3$ molecules are labile, which can be substituted by other phosphines such as PCy$_3$, P(OEt)Ph$_2$, and P(OEt)$_3$ to give (η^2-$C_2B_{10}H_{10}$)Ni(PPh$_3$)(PCy$_3$), (η^2-$C_2B_{10}H_{10}$)Ni[P(OEt)Ph$_2$]$_2$, and (η^2-$C_2B_{10}H_{10}$)Ni[P(OEt)$_3$]$_2$, respectively with quantitative conversion.

Our earlier investigation on the B-substituted carborane has revealed that the substituent on carborane plays an important role for the formation of these late transition metal complexes. The reactions of $Li_2C_2B_{10}Me_8H_2$ with (PPh$_3$)$_2$NiCl$_2$ or (dppe)NiCl$_2$ gave redox reaction products (σ-$C_2B_{10}Me_8H_3$)Ni(PPh$_3$)$_2$ or (η^2-$C_2B_{10}Me_8H_2$)Ni(μ-σ:σ:η^2-dppen)Ni(dppe), whereas the complete redox reaction took place for the reactions with (PPh$_3$)$_2$PdCl$_2$ and (PPh$_3$)$_2$PtCl$_2$. In view of the very rich and exciting chemistry of nickel-benzyne complexes [15–19], we are interested in further exploring the nickel-1,2-*o*-carboryne complexes with

Z. Qiu, *Late Transition Metal-Carboryne Complexes*, Springer Theses,
DOI: 10.1007/978-3-642-24361-5_2, © Springer-Verlag Berlin Heidelberg 2012

substituents on the cage boron. In this section we will describe the synthesis and structure of these B-substituted nickel-1,2-o-carboryne complexes.

2.2 Synthesis and Structure of B-Substituted Nickel-1,2-o-Carboryne Complexes

Treatment of 9-I-1,2-C$_2$B$_{10}$H$_{11}$ and 9,12-I$_2$-1,2-C$_2$B$_{10}$H$_{11}$ with 2 equiv of n-BuLi in THF at 0 °C, followed by reaction with 1 equiv of (PPh$_3$)$_2$NiCl$_2$ in THF at the temperatures −30 °C to room temperature gave (η^2-9-I-1,2-C$_2$B$_{10}$H$_9$)Ni(PPh$_3$)$_2$ (**II-1**) as a yellow solid or (η^2-9,12-I$_2$-1,2-C$_2$B$_{10}$H$_8$)Ni(PPh$_3$)$_2$ (**II-2**) as yellow crystals in 55 or 72% isolated yield, respectively (Scheme 2.1).

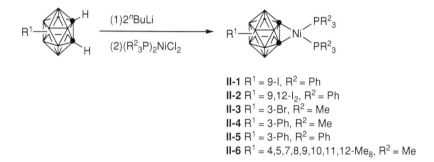

II-1 R^1 = 9-I, R^2 = Ph
II-2 R^1 = 9,12-I$_2$, R^2 = Ph
II-3 R^1 = 3-Br, R^2 = Me
II-4 R^1 = 3-Ph, R^2 = Me
II-5 R^1 = 3-Ph, R^2 = Ph
II-6 R^1 = 4,5,7,8,9,10,11,12-Me$_8$, R^2 = Me

Scheme 2.1 Synthesis of B-substituted 1,2-o-Carboryne-Ni complexes

Under the same conditions, the product from the reaction of 3-bromo-1,2-dilithio-o-carborane with 1 equiv of (PPh$_3$)$_2$NiCl$_2$ was not stable and decomposed to generate 3-Br-1,2-C$_2$B$_{10}$H$_{11}$ and Ni(0) species. However, the less bulky and more electron-donating ligand of PMe$_3$ can stabilize this Ni-1,2-o-carboryne complex and (η^2-3-Br-1,2-C$_2$B$_{10}$H$_9$)Ni(PMe$_3$)$_2$ (**II-3**) can be synthesized as yellow crystals in 31% isolated yield from the interaction of 3-bromo-1,2-dilithio-o-carborane with 1 equiv of (Me$_3$P)$_2$NiCl$_2$ in THF at the temperatures −30 °C to room temperature (Scheme 2.1).

Similarly, complexes (η^2-3-C$_6$H$_5$-1,2-C$_2$B$_{10}$H$_9$)Ni(PMe$_3$)$_2$ (**II-4**) and (η^2-3-C$_6$H$_5$-1,2-C$_2$B$_{10}$H$_9$)Ni(PPh$_3$)$_2$ (**II-5**) can be isolated as yellow or orange crystals in 42 or 76% isolated yields by the reaction of 3-phenyl-1,2-dilithio-o-carborane with 1 equiv of (Me$_3$P)$_2$NiCl$_2$ or (Ph$_3$P)$_2$NiCl$_2$ in THF, respectively.

In the case of 4,5,7,8,9,10,11,12-octamethyl-o-carborane, neither PPh$_3$ nor PMe$_3$ can efficiently stabilize the corresponding Ni-1,2-o-carboryne species, leading to a

2.2 Synthesis and Structure of B-Substituted Nickel-1,2-o-Carboryne Complexes

mixture of products. A few brown X-ray-quality-crystals of (η^2-4,5,7,8,9,10,11, 12-Me$_8$-C$_2$B$_{10}$H$_2$)Ni(PMe$_3$)$_2$ (**II-6**) was obtained from toluene solution at room temperature during the recrystallization of the product.

Complexes **II-1–4** are sensitive to moisture and air whereas **II-5** is air- and moisture-stable, both in the solid-state and in solution. Interestingly, **II-5** is very thermally stable even in refluxing THF. However, heating the THF solution of **II-1–4** can lead to a decomposition, generating neutral carboranes and Ni(0) species. It's believed that both the interaction between phenyl ring and metal center and the sterically demanding PPh$_3$ ligand have contributed to the exceptional stability of this complex. Complexes **II-1–5** are slightly soluble in ether and toluene and highly soluble in THF. They were fully characterized by various spectrometric methods and elemental analyses.

The ^{1}H and ^{13}C NMR spectra of **II-1–5**, which only display the signals of PPh$_3$ or PMe$_3$ ligand, do not give much information on the solution structures. The cage carbons were not observed for **II-1–5**. The ^{11}B{^{1}H} NMR spectra display a 3:6:1, 2:6:2, 1:1:2:2:2:2, 2:1:3:4, and 1:1:1:2:5 pattern for **II-1**, **II-2**, **II-3**, **II-4**, and **II-5** respectively. One singlet at $\sim$22 ppm corresponding to BI vertex can be observed for **II-1** and **II-2**. The BBr signal of **II-3** is overlapped with other BH signals at -10 to -14 ppm. In the ^{11}B NMR spectra of **II-4** and **II-5**, the BPh signal can be observed as a singlet at -3 ppm. The ^{31}P NMR spectra show one singlet at $\sim$30 ppm for PPh$_3$ ligand in **II-1,2,5** or one singlet at ~ -9 ppm for PMe$_3$ ligand in **II-3,4**.

The B–H$\cdots$M interactions in late transition-metal complexes usually lead to a significantly reduced J_{BH} value, a very deshielded ^{11}B signal, and a very high-field ^{1}H resonance [20–24]. It is noted that there is no significant high-field ^{1}H signal and low-field ^{11}B signal observed in the ^{1}H and ^{11}B NMR spectra of **II-1–5**. The IR spectra (KBr) exhibited one very strong and broad stretching band v_{B-H} at about 2570 cm^{-1} for **II-1–3**, whereas that of **II-4** and **II-5** showed two v_{B-H} bands at 2550 and 2530 cm^{-1} (**II-4**)/2510 cm^{-1} (**II-5**).

The solid-state structures of **II-2–6** were further confirmed by single-crystal X-ray analyses. As shown in Figs. 2.1, 2.2, 2.3, 2.4, 2.5, complexes **II-2–6** have similar coordination geometries, which contain a three-membered ring formed through two Ni–C(cage) bonds and the coordination plane about the nickel atom is essentially planar. There are two crystallographically independent molecules in the unit cell of **II-5**. Figure 2.4 shows its representative structure. Selected bond distances and angles around the metal centers are listed in Table 2.1 for comparison.

The C(cage)–C(cage) bond distances (1.55–1.57 Å) are close to each other for **II-2,3,4,6** and similar with those observed in the Ni-1,2-o-carboryne complexes [1.556(5) Å in (η^2-C$_2$B$_{10}$H$_{10}$)Ni(PPh$_3$)$_2$, 1.576(6) Å in (η^2-C$_2$B$_{10}$Me$_8$H$_2$)Ni(μ-σ:σ:η^2-dppen)Ni(dppe), 1.551(4) Å in (η^2-C$_2$B$_{10}$H$_{10}$)Ni(PPh$_3$)(PCy$_3$), 1.553(6) Å and 1.561(5) Å in (η^2-C$_2$B$_{10}$H$_{10}$)Ni(dppe)], which is much shorter than that of 1.63 Å found in o-carborane. The large steric effect results in the shorter C(cage)–C(cage) bond distances of 1.523(3) Å in **II-5**. It is noteworthy that, since all B–H hydrogen atoms are in calculated positions, a detailed discussion of the B–H distances is not warranted.

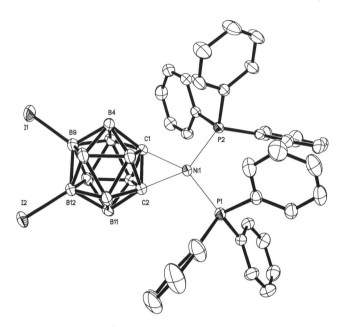

Fig. 2.1 Molecular structure of $(\eta^2\text{-}9,12\text{-}I_2\text{-}1,2\text{-}C_2B_{10}H_8)Ni(PPh_3)_2$ (**II-2**)

Fig. 2.2 Molecular structure of $(\eta^2\text{-}3\text{-}Br\text{-}1,2\text{-}C_2B_{10}H_9)Ni(PPh_3)_2$ (**II-3**)

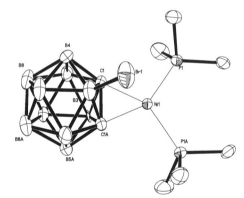

Fig. 2.3 Molecular structure of $(\eta^2\text{-}3\text{-}C_6H_5\text{-}1,2\text{-}C_2B_{10}H_9)Ni(PMe_3)_2$ (**II-4**)

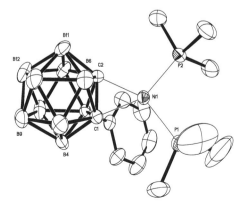

2.2 Synthesis and Structure of B-Substituted Nickel-1,2-o-Carboryne Complexes 35

Fig. 2.4 Molecular structure of (η^2-3-C$_6$H$_5$-1,2-C$_2$B$_{10}$H$_9$)Ni(PPh$_3$)$_2$ (**II-5**), showing one of two independent molecules in the unit cell

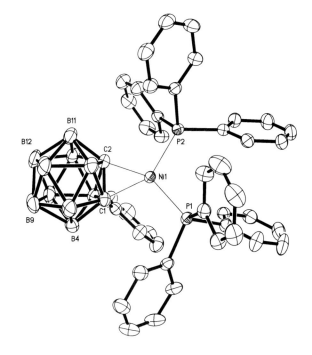

Fig. 2.5 Molecular structure of (η^2-4, 5, 7, 8, 9, 10, 11, 12-Me$_8$-1,2-C$_2$B$_{10}$H$_2$)Ni(PMe$_3$)$_2$ (**II-6**)

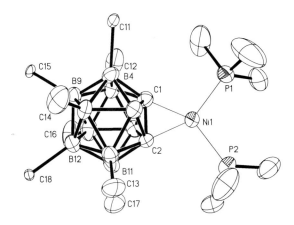

Table 2.1 Selected bond distances (Å) and angles (deg) for late transition metal-1,2-o-carboryne complexes

	II-2	II-3	II-4		II-5	II-6
C(1)-C(2)	1.550(6)	1.595(14)	1.565(5)	1.576(4)	1.523(3)	1.562(14)
Ni(1)-C(1)	1.926(4)	1.918(6)	1.917(3)	1.927(3)	1.950(2)	1.923(10)
Ni(1)-C(2)	1.917(4)	1.918(6)	1.929(3)	1.924(3)	1.924(2)	1.899(10)
Ni(1)···B(3)	2.647(6)					2.613(12)
Ni(1)···B(6)	2.626(6)	2.721(20)	2.631(4)	2.583(4)	2.581(3)	2.609(11)
Ni(1)-P(1)	2.190(1)	2.169(2)	2.166(1)	2.164(1)	2.215(1)	2.166(3)
Ni(1)-P(2)	2.214(1)	2.169(2)	2.169(1)	2.163(1)	2.221(1)	2.167(3)
Ni(1)-C(1)-C(2)	65.9(2)	65.4(2)	66.3(2)	65.7(2)	65.9(1)	65.1(5)
Ni(1)-C(2)-C(1)	66.5(2)	65.4(2)	65.6(2)	65.9(2)	67.7(1)	66.7(5)
C(1)-Ni(1)-C(2)	47.6(2)	49.1(4)	48.1(1)	48.3(1)	46.3(1)	48.2(4)

2.3 Summary

We have prepared several B-substituted nickel-1,2-o-carboryne complexes and fully characterized their structures. Our studies show that the reaction between phosphine ligated metal halide and $Li_2C_2B_{10}H_{10}$ is a good synthetic route for the preparation of these complexes. These late transition metal-1,2-o-carboryne complexes have similar structural features and the C(cage)–C(cage) bonds in these complexes are much shorter than those in Zr-1,2-o-carboryne complexes.

References

1. Gingrich HL, Ghosh T, Huang Q, M Jones Jr (1990) J Am Chem Soc 112:4082
2. Kiran B, Anoop A, Jemmis ED (2002) J Am Chem Soc 124:4402
3. Huang Q, Gingrich HL, Jones M Jr (1991) Inorg Chem 30:3254
4. Cunningham RT, Bian N, Jones M Jr (1994) Inorg Chem 33:4811
5. Ho DM, Cunningham RJ, Brewer JA, Bian N, Jones M Jr (1995) Inorg Chem 34:5274
6. Barnett-Thamattoor L, Zheng G, Ho DM, Jones M Jr, Jackson JE (1996) Inorg Chem 35:7311
7. Atkins JH, Ho DM, Jones M Jr (1996) Tetrahedron Lett 37:7217
8. Lee T, Jeon J, Song KH, Jung I, Baik C, Park KM, Lee SS, Kang SO, Ko J (2004) Dalton Trans, p 933
9. Jeon J, Kitamura T, Yoo B-W, Kang SO, Ko J (2001) Chem Commun, p 2110
10. Ren S, Deng L, Chan H-S, Xie Z (2009) Organometallics 28:5749
11. Wang H, Li H-W, Huang X, Lin Z, Xie Z (2003) Angew Chem Int Ed 42:4347
12. Sayler AA, Beall H, Sieckhaus JF (1973) J Am Chem Soc 95:5790
13. Zakharkin LI, Kovredov AI (1975) Izv Akad Nauk SSSR Ser Khim, p 2619
14. YuA Ol'dekop, Maier NA, Erdman AA, Prokopovich VP (1982) Zh Obshch Khim 52:2256
15. Buchwald SL, Nielsen RB (1988) Chem Rev 88:047
16. Bennett MA, Schwemlein HP (1989) Angew Chem Int Ed Engl 28:1296
17. Bennett MA, Wenger E (1997) Chem Ber 130:1029
18. Jones WM, Klosin J (1998) Adv Organomet Chem 42:147

References

19. Retbøll M, Edwards AJ, Rae AD, Willis AC, Bennett MA, Wenger E (2002) J Am Chem Soc 124:8348
20. Brew SA, Stone FGA (1993) Adv Organomet Chem 35:135
21. Franken A, McGrath TD, Stone FGA (2006) J Am Chem Soc 128:16169
22. McGrath TD, Du S, Hodson BE, Lu XL, Stone FGA (2006) Organometallics 25:4444
23. McGrath TD, Du S, Hodson BE, Stone FGA (2006) Organometallics 25:4452
24. Hodson BE, McGrath TD, Stone FGA (2005) Organometallics 24:3386

Chapter 3
Nickel-Mediated Coupling Reaction of 1,2-o-Carboryne with Alkenes

3.1 Introduction

Metal-benzyne complexes have found many applications in organic synthesis, mechanistic studies, and the synthesis of functional materials [1–5]. In contrast, their analogues, metal-1,2-o-carboryne complexes are largely unexplored although the reactivity pattern of 1,2-o-carboryne (generated in situ) has been actively investigated [6–13].

Cp$_2$Zr(η^2-C$_2$B$_{10}$H$_{10}$) [produced in situ from the precursor of Cp$_2$Zr(μ-Cl) (μ-C$_2$B$_{10}$H$_{10}$)Li(OEt$_2$)$_2$] has a similar reactivity pattern to that of Cp$_2$Zr(η^2-C$_6$H$_4$) in reactions with polar and nonpolar unsaturated organic substrates [14, 15]. It reacts well with isonitrile, nitrile, azide, alkene, and alkyne to give monoinsertion products. On the other hand, (η^2-C$_2$B$_{10}$H$_{10}$)Ni(PPh$_3$)$_2$ (**III-1**) undergoes regioselective [2+2+2] cycloaddition with alkynes affording benzocarboranes in a head-to-tail manner, but it does not react with the aforementioned polar unsaturated molecules [16]. These results indicate that the nature of transition metals plays a crucial role in these reactions.

In view of the reactions of nickel-benzynes with alkene to generate monoinsertion products (Scheme 3.1) [17] and the analogy between nickel-benzyne and nickel-1,2-o-carboryne complexes (Chart 3.1), we are interested in exploring the reactivity of nickel-1,2-o-carboryne with alkenes. In this section we will describe the reaction of nickel-1,2-o-carboryne with alkenes affording alkenylcarboranes.

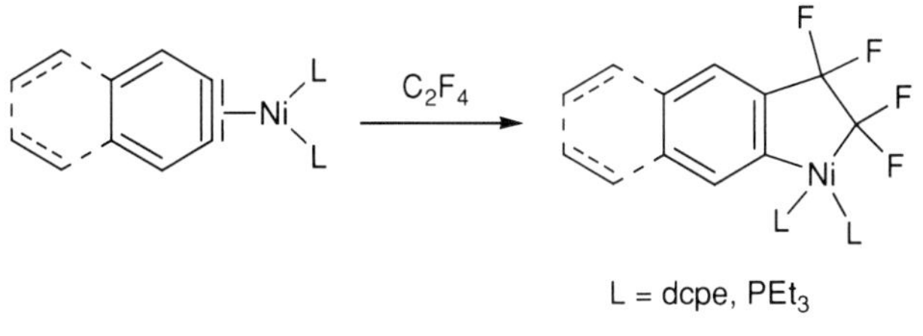

$$L = \text{dcpe, PEt}_3$$

Scheme 3.1 Reaction of Ni-benzyne with alkene

Z. Qiu, *Late Transition Metal-Carboryne Complexes*, Springer Theses,
DOI: 10.1007/978-3-642-24361-5_3, © Springer-Verlag Berlin Heidelberg 2012

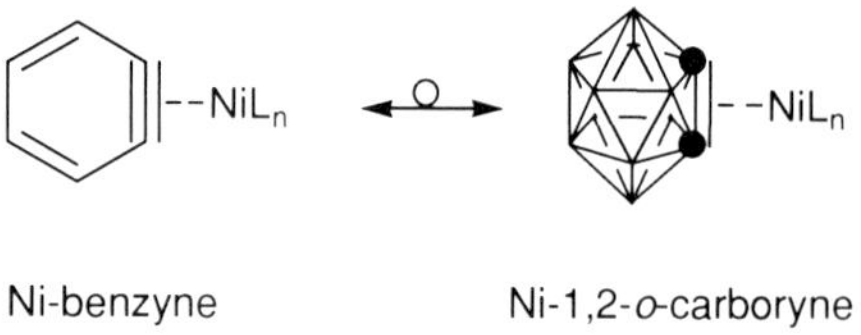

Chart 3.1 Isolobal analogue

3.2 Results and Discussion

A typical procedure is as follows. To a THF solution (10 mL) of $Li_2C_2B_{10}H_{10}$ (1.0 mmol), prepared in situ from the reaction of ^nBuLi (2.0 mmol) with o-carborane (1.0 mmol), was added $(PPh_3)_2NiCl_2$ (1.0 mmol) at 0 °C. The reaction mixture was further stirred for 0.5 h at room temperature giving the Ni-1,2-o-carboryne intermediate $(\eta^2$-$C_2B_{10}H_{10})Ni(PPh_3)_2$ (**III-1**) [16, 18]. Alkene (2 equiv) was added at room temperature, and the reaction mixture was heated at 90 °C in a closed vessel overnight. The reaction mixture was then cooled to room temperature and quenched with $NaHCO_3$ solution. Normal workup afforded the coupling products in excellent regio- and steroselectivity for most alkene as shown in Table 3.1.

The temperature is crucial for this reaction. No reaction proceeded at $T < 60$ °C. On the other hand, higher reaction temperatures (>90 °C) led to the decomposition of **III-1** as indicated by ^{11}B NMR. Toluene and diethyl ether were not suitable for this reaction because of the poor solubility of **III-1**. Other phosphines such as PEt_3, $P(OEt)_3$, and dppe gave very similar results to that of PPh_3. It is noted that the same results were obtained if the isolated pure complex **III-1** was used for the reactions.

As shown in Table 3.1, a variety of alkenes is compatible with this nickel-mediated cross-coupling reaction. Substituted styrenes reacted efficiently to give "Heck type" of products **III-3** as single regioisomers with excellent stereoselectivity in very good isolated yields. The nature of the substituents on phenyl ring has no obvious effect on the reaction results (entries 1–5). The yields were lower for 1,1-diphenylethene and vinyltrimethylsilane due to steric effects (entries 7 and 8). The "ene-reaction type" of products was isolated in good yields for aliphatic alkenes and α-methylstyrene (**III-2f**) (entries 6, 9–11). For example, **III-4k** was isolated in 67% yield, which is much higher than the 10–20% yield from the direct reaction of 1,2-o-carboryne with cyclohexene [7–11]. Vinylethers **III-2n** and **III-2o** also reacted with Ni-1,2-o-carboryne **III-1** but to a less extent probably due to the coordination of oxygen atom occupying the vacant site of the Ni atom (entries 14 and 15). Such interactions may alter the regioselectivity of the olefin insertion and stabilize the inserted product, leading to the formation of **III-5n** after hydrolysis. In the case of norbornene (**III-2l**), the corresponding inserted product was thermodynamically very stable, [19, 20] affording only hydrolysis product

3.2 Results and Discussion

Table 3.1 Nickel-mediated coupling reaction of 1,2-*o*-carboryne with alkene

Entry	Alkene	Product	Yield (%)[a]
1	III-2a	III-3a	82
2	III-2b	III-3b	85
3	III-2c	III-3c	80
4	III-2d	III-3d	73
5	III-2e	III-3e	76
6	III-2f	III-4f	59

(continued)

Table 3.1 (continued)

Entry	Alkene	Product	Yield (%)[a]
7	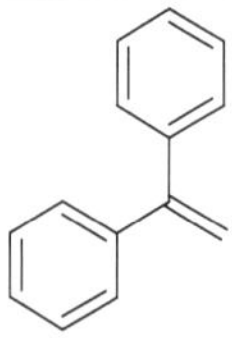 III-2g	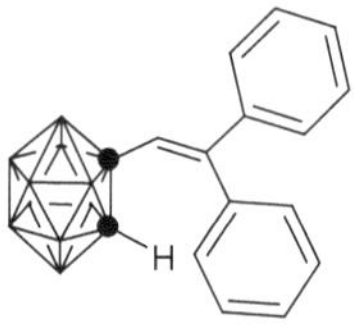 III-3g	46
8	 III-2h	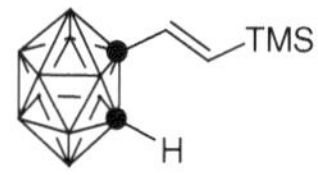 III-3h	46
9	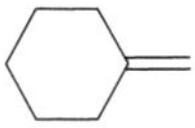 III-2i	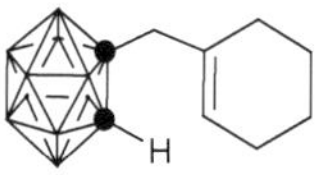 III-4i	77
10	 III-2j	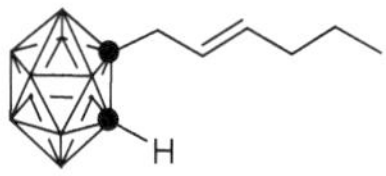 III-4j 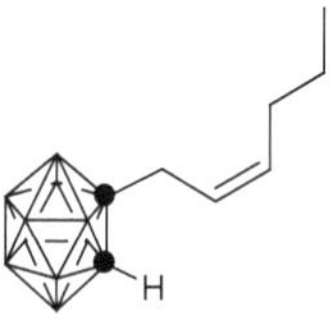 III-5j	74 (1:1)[b]
11	 III-2k	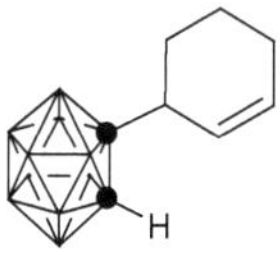 III-4k	67
12	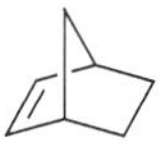 III-2l	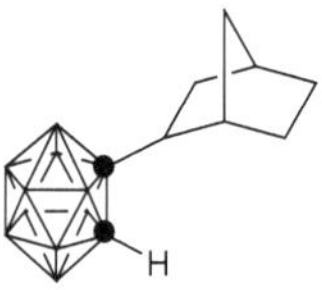 III-5l	60[c]

(continued)

3.2 Results and Discussion

Table 3.1 (continued)

Entry	Alkene	Product	Yield (%)[a]
13	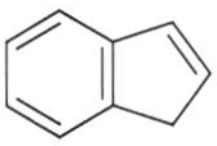 III-2m	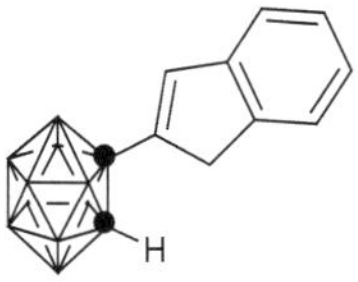 III-3m	31
		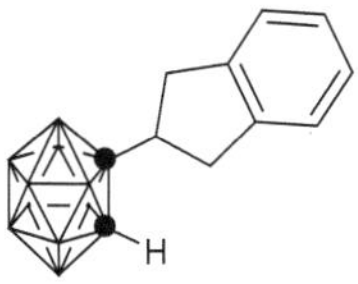 III-5m	27[c]
14	 III-2n	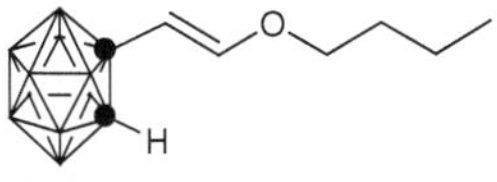 III-3n	18
		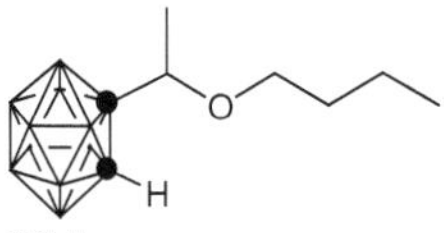 III-5n	12[c]
15	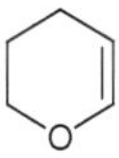 III-2o	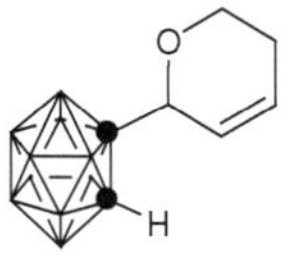 III-4o	15

[a] Isolated yields

[b] **III-4j/5j** were inseparable and their ratio was estimated by [1] H NMR

[c] Isolated after hydrolysis

III-5l in 60% isolated yield (entry 12). No double insertion product was observed. For indene (**III-2m**), both hydrolysis product **III-5m** and "ene-reaction type" of species **III-3m** were isolated in 27 and 31% yield, respectively (entry 13). No reaction was proceeded with *cis-* and *trans-*stillbene, 6,6-dimethylfulvene, 1,1-dimethylallene, 1-phenylallene, 2-propenenitrile, diphenylvinylphosphine, ethylvinylsulfide, anthracene, furan, and thiophene.

All new products were fully characterized by various spectroscopic techniques and high resolution MS. In the ^{1}H NMR spectra of **III-3a–e, h**, and **n**, ^3J ($\sim$15 Hz) for olefinic protons indicates a *trans-*conformation. In the reaction of 1-hexene, *trans-*and *cis-*isomers were observed with the molar ratio of 1:1 in the ^{1}H NMR spectrum of the crude product. The ^{11}B NMR spectra generally exhibited a 1:1:2:2:2:2 splitting pattern for all these mono-substituted carboranes.

The molecular structures of **III-3c, III-3e**, and **III-3g** were further confirmed by single-crystal X-ray analyses (Figs. 3.1, 3.2, 3.3).

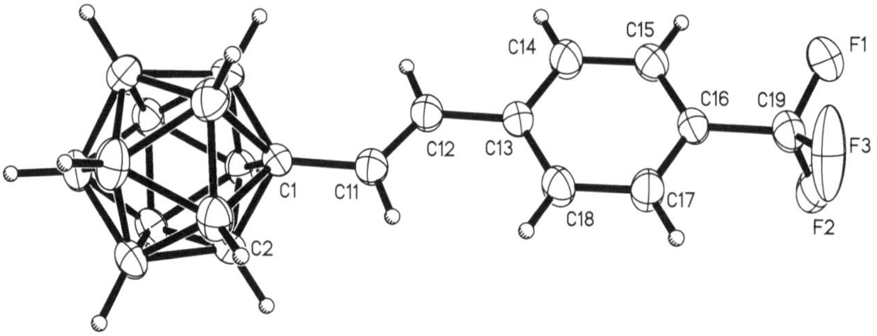

Fig. 3.1 Molecular structure of **III-3c**

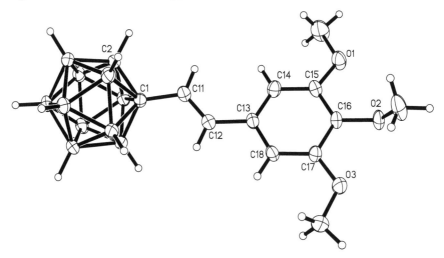

Fig. 3.2 Molecular structure of **III-3e**

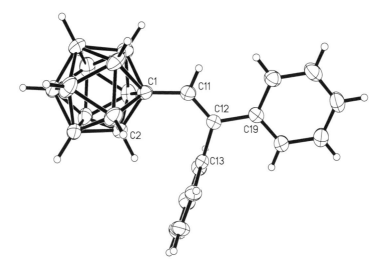

Fig. 3.3 Molecular structure of **III-3g**

3.2 Results and Discussion 45

It has been documented that the reaction of 1,2-*o*-carboryne (generated in situ) with anthracene, furan, or thiophene gave [2+4] cycloaddition products [6–13]. The Ni-1,2-*o*-carboryne complex **III-1**, however, did not react with any of them. This result suggests that 1,2-*o*-carboryne and Ni-1,2-*o*-carboryne should undergo different reaction pathways in the reactions with alkene.

Scheme 3.2 shows the plausible mechanism for the formation of coupling products. Dilithiocarborane reacts with $(PPh_3)_2NiCl_2$ to generate the Ni-1,2-*o*-carboryne complex **III-1** [18]. Coordination and insertion of alkene give a nickelacycle **III-A** [21–23]. The regioselectivity observed in the reaction can be rationalized by the large steric effect of carborane moiety. β-H/β'-H elimination prior to the insertion of the second molecule of alkene produces the intermediate **III-B/B'** [24–27]. Reductive elimination affords the alkenylcarboranes **III-3** ("Heck type" of products) or **III-4** ("ene-reaction type" of products). In general, β-H elimination of five-membered metallacycles is more difficult than β'-H elimination (vide infra) [24–27]. Such hydrogen elimination reactions may be suppressed due to steric reasons [19] or intramolecular coordination of the heteroatom, which leads to the formation of alkylcarboranes after hydrolysis (Table 3.1, entries 12–14).

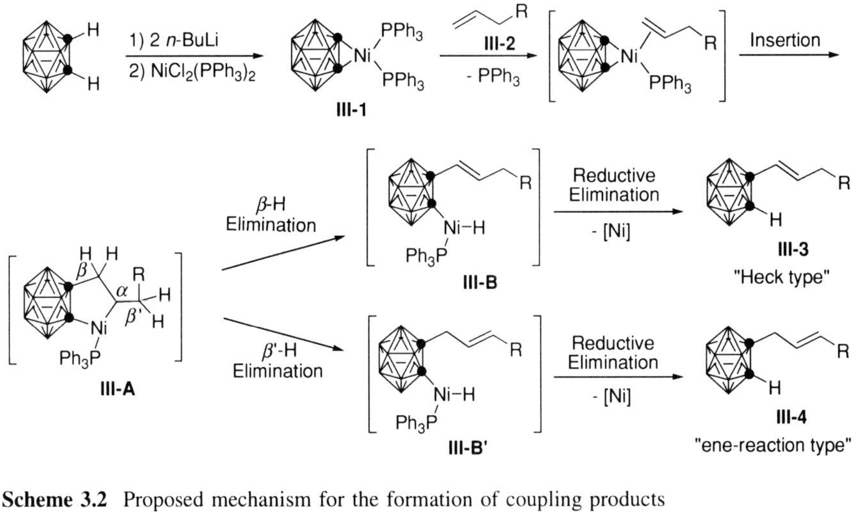

Scheme 3.2 Proposed mechanism for the formation of coupling products

The aforementioned mechanism is supported by the following experiment. Treatment of **III-1** with styrene-d_3 in THF at 90 °C gave $[D_3]$-**III-3a** in 80% isolated yield with >95% deuterium incorporation (Scheme 3.3).

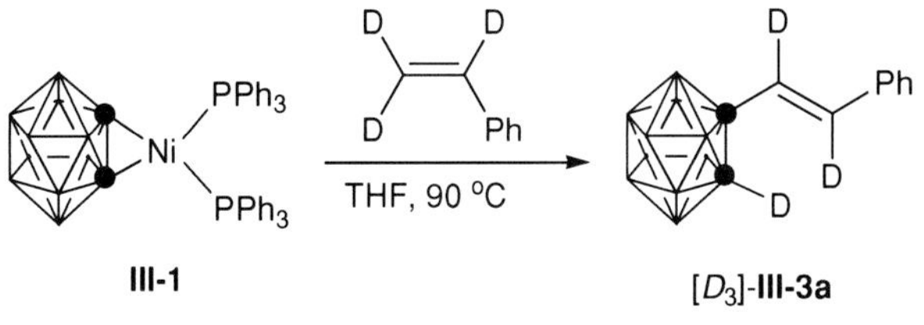

Scheme 3.3 Reaction of **III-1** with styrene-d_3

It has been suggested that many metallacycles, such as six-, five-, four-, and three-membered metallacycles, cannot undergo β-H elimination readily, because the $M\text{-}C_{\alpha}\text{-}C_{\beta}$-H dihedral angles in these compounds are constrained to values far from $0°$. In addition, the β-hydrogen atoms are constrained to positions far away from the metal center. These theories are conformed to our result that β-H elimination of five-membered metallacycles is more difficult than β'-H elimination. However, there is increasing evidence that β-H elimination reactions of five-membered-ring intermediates to afford hydridometal alkene complexes are possible (Scheme 3.4) [24–27].

Scheme 3.4 β-H elimination

It is well-known that cyclopentane actually assumes a slightly puckered "envelope" conformation that reduces the eclipsing and lowers the torsional strain. This puckered shape is not fixed but undulates by the thermal up-and-down motion of the five methylene groups (Scheme 3.5, **I**). In the metallacycle, there are three

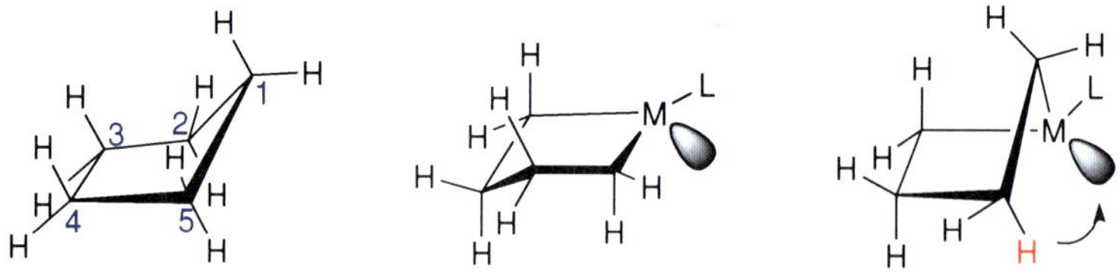

Scheme 3.5 β-H elimination of five-membered metallacycle

types of unique positions in the five-membered ring. In consideration of the steric effect of the ligand, the metal atom can only be located at two positions (Scheme 3.5, **II** and **III**). The two conformational isomers are both 14 electron species. Because of their coordinatively unsaturated feature, each isomer is expected to have a low-lying unoccupied orbital, which should have the maximum amplitude along the direction of the missing leg in the four-legged piano-stool structure. The low-lying unoccupied orbital is ready to accept the β-hydrogen to form the metal-hydride bond in the eliminated product. It can be seen that in **II** all the β-hydrogens orient themselves away from the maximum amplitude of the low-lying unoccupied orbital. However, there is a β-hydrogen in close proximity to the maximum amplitude of the low-lying unoccupied orbital in **III**. The close proximity of the transferring β-hydrogen to the accepting unoccupied orbital is essential to facilitate the β-hydrogen elimination process [27].

In the reaction of indene (**III-2m**), due to the steric reason, indene prefers to insert as showed in pathway b to give intermediate **III-D** than afford **III-4m** through intermediate **III-C** in pathway a (Scheme 3.6). After quenching, the

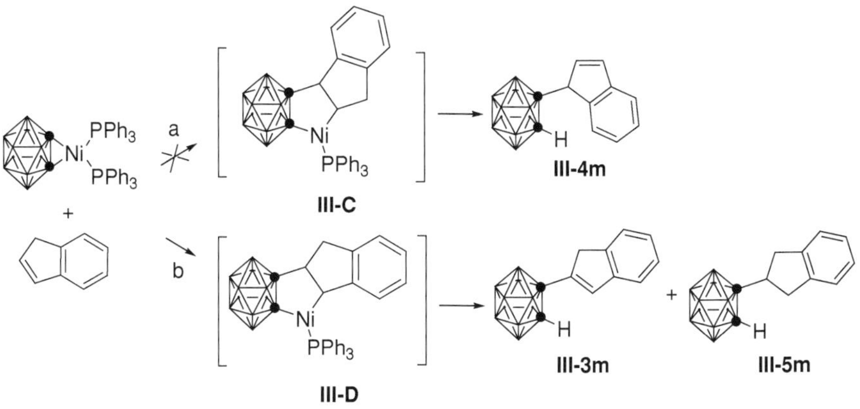

3.2 Results and Discussion   47

Scheme 3.6 Reaction of **III-1** with indene

hydrolysis product **III-5m** without β-H elimination is observed along with the normal product **III-3m** in 0.9:1 molar ratio. It can be considered that there are two β-H conformations in **III-D** (*exo* and *endo* to Ni). The β-H$_{endo}$ elimination affords **III-3m**. Due to the fused ring structure, the coplanar conformation is difficult to be achieved by β-H$_{exo}$ and Ni atom, and the hydrolysis species **III-5m** is obtained as the final product.

In the reaction of norbornene (**III-2l**), the metallacycle **III-E** is especially stable owing to the absence of β-H atom with appropriate geometric requirements for elimination, the alkyl-substituted product was obtained after quenching (Scheme 3.7) [19, 20].

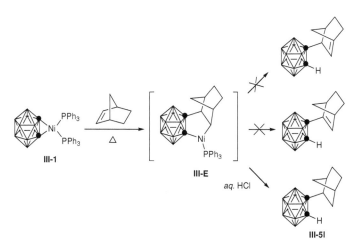

Scheme 3.7 Reaction of **III-1** with norbornene

The reactions of nickel-1,2-*o*-carboryne **III-1** with alkenes having donor atom such as methyl acrylate and 2-vinylpyridine were also investigated (Schemes 3.8 and 3.10). The coordination of these donor atoms in olefin may stabilize the intermediate, preventing the β-H elimination.

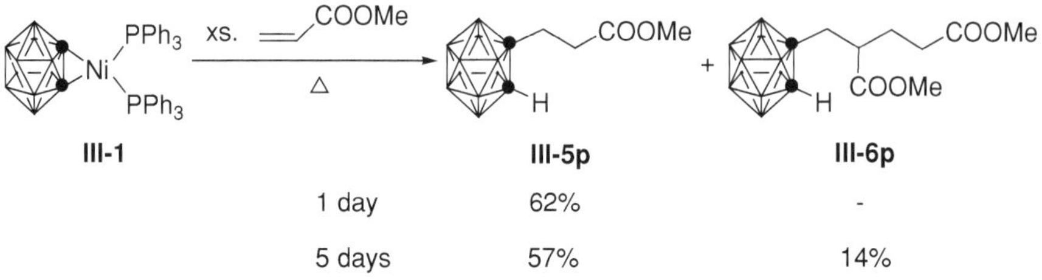

| | 1 day | 62% | - |
| | 5 days | 57% | 14% |

Scheme 3.8 Reaction of **III-1** with methyl acrylate

In case of methyl acrylate, the absence of alkenylcarborane can be rationalized by the coordination of the carbonyl to the Ni atom, stabilizing the nickelacycle intermediate. Product **III-6p**, generated from the second molecule of methyl acrylate insertion into the Ni–C$_{alkyl}$ bond, was obtained by extending the reaction time to 5 days.

To investigate the reaction mechanism, the following labeling experiment was performed. Quenching the reaction mixture with D$_2$O can afford the desired product [*D$_2$*]**III-5p** with greater than 95% deuterium incorporation (Scheme 3.9).

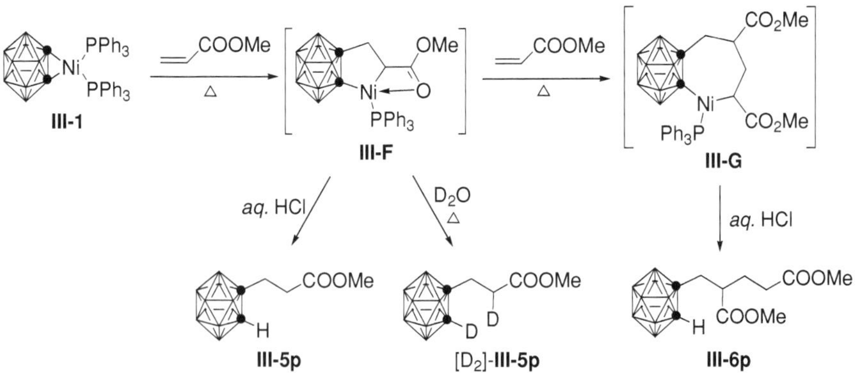

Scheme 3.9 Proposed mechanism for the formation of alkylcaborane product

The reaction of 2-vinylpyridine is different from that of methyl acrylate. Although similar mono-alkylcarborane was obtained after heating at 90 °C overnight and

3.2 Results and Discussion

quenching, extension of reaction time can lead to the isolation of two new products, **III-7q** and **III-8q**, which should be generated by the insertion of the second molecule of 2-vinylpyridine into the Ni–C(cage) bond (Scheme 3.10). It's a very rare example for a M–C(cage) bond to be involved in the reactions because the unique electronic and steric properties of carboranyl moiety can make the M–C(cage) bond in metal-carboranyl complexes inert toward unsaturated molecules [28–31].

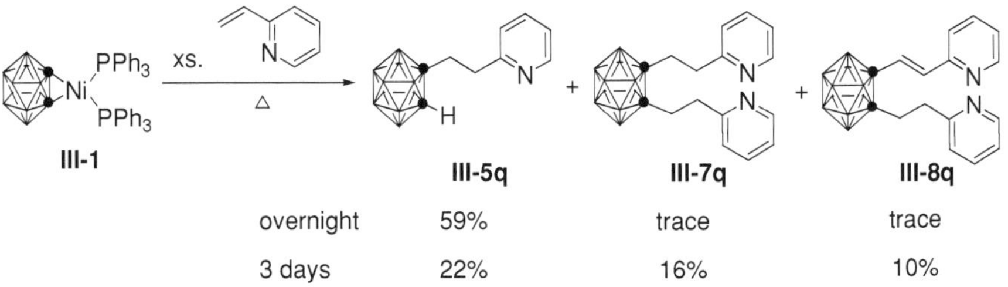

	III-5q	III-7q	III-8q
overnight	59%	trace	trace
3 days	22%	16%	10%

Scheme 3.10 Reaction of **III-1** with 2-vinylpyridine

The ^{1}H NMR spectra of **III-5q** and **III-8q** are compared with that of **III-7q** (Fig. 3.4). In the ^{1}H NMR spectrum of **III-5q**, one singlet at 3.80 ppm corresponding to the cage C*H* can be clearly observed. The ^{1}H NMR spectrum of **III-8q** clearly showed two doublets at 6.97 and 7.03 ppm with $^3J = 15.3$ ppm corresponding to *trans*-olefinic protons. And two sets of pyridinyl signals indicate the unsymmetrical structure of **III-8q**.

The molecular structures of **III-7q**, **III-8q**, and **III-5p** were further confirmed by single-crystal X-ray analyses as shown in Figs. 3.5, 3.6 and 3.7 (Table 3.2).

Scheme 3.11 shows the plausible mechanism for the formation of **III-7q** and **III-8q**. A second equiv of 2-vinylpyridine can insert into the Ni–C(cage) bond in nickelacyclopropane **III-H** to afford nickelacycloheptane **III-I**, which can undergo β-H elimination to give Ni–H species **III-J**. Quenching of these coexisting intermediates of **III-H**, **III-I**, and **III-J** in the reaction mixture leads to the formation of **III-5q**, **III-7q**, and **III-8q**, respectively.

The mono-alkene insertion species nickelacyclopentanes **III-9p,q** (Chart 3.2) were isolated and fully characterized from the reaction of **III-1** with methyl acrylate and 2-vinylpyridine, respectively. Complex **III-9q** was further confirmed by single-crystal X-ray analyses. It is an ionic complex, in which the anion consists of three square-planar Ni moieties sharing one μ_3-Cl atom (Fig. 3.8). The proposed molecular structure of **III-9p** is shown in Chart 3.2, which is supported by ^{1}H, ^{13}C and ^{11}B NMR as well as elemental analyses.

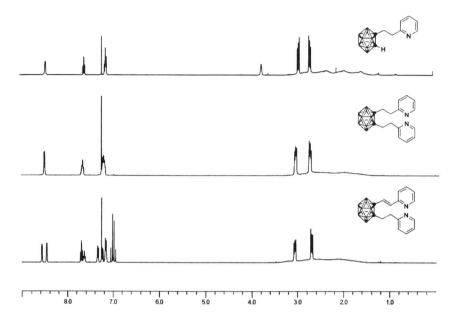

Fig. 3.4 ¹H NMR spectra of **III-5q**, **III-7q**, and **III-8q**

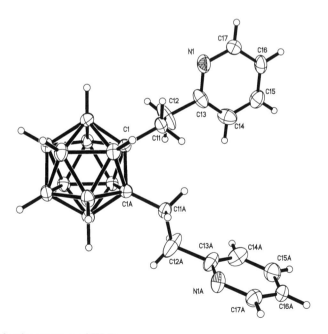

Fig. 3.5 Molecular structure of **III-7q**

3.2 Results and Discussion

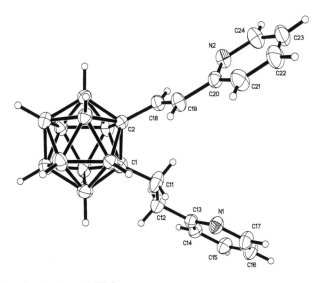

Fig. 3.6 Molecular structure of **III-8q**

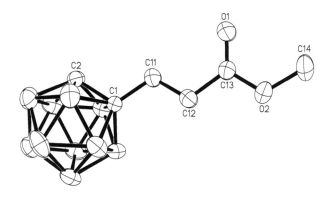

Fig. 3.7 Molecular structure of **III-5p**

Table 3.2 Selected bond distances (Å) and angles (deg) for **III-7q** and **III-8q**

III-7q		III-8q			
C1-C11	1.522(3)	C1-C11	1.513(3)	C2-C18	1.497(2)
C11-C12	1.482(4)	C11-C12	1.406(3)	C18-C19	1.347(3)
C12-C13	1.510(3)	C12-C13	1.492(3)	C19-C20	1.474(3)
C1-C11-C12	116.2(2)	C1-C11-C12	118.4(2)	C2-C18-C19	123.3(2)
C11-C12-C13	112.2(2)	C11-C12-C13	117.8(2)	C18-C19-C20	122.6(2)

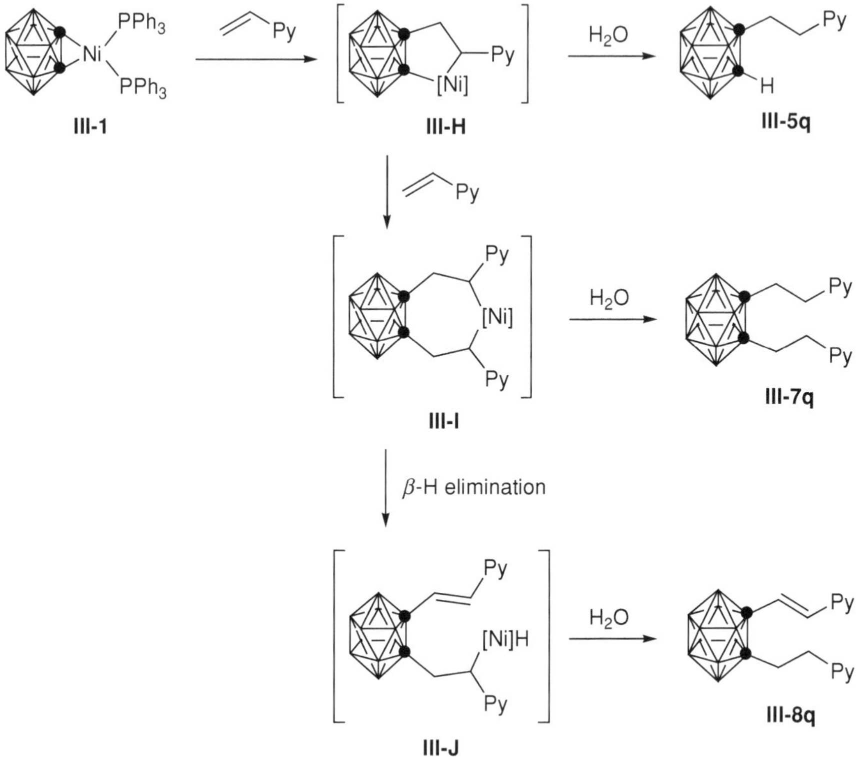

52   3 Nickel-Mediated Coupling Reaction of 1,2-*o*-Carboryne with Alkenes

Scheme 3.11 Proposed mechanism for the formation of **III-7q** and **III-8q**

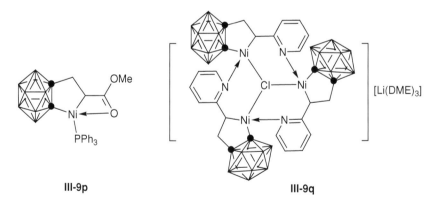

Chart 3.2 Structure of **III-9p** and **III-9q**

3.3 Summary

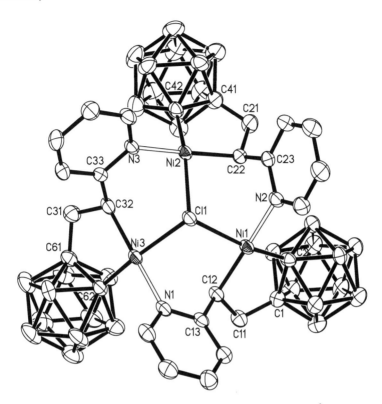

Fig. 3.8 Molecular structure of the anion in **III-9q**. Selected bond lengths (Å) and angles (deg): Ni1-C2 1.884(5), Ni1-C12 1.966 (5), Ni1-Cl1 2.292(1), Ni1-N2 1.937(4), Ni2-C42 1.884(6), Ni2-C22 1.970(5), Ni2-Cl1 2.307(1), Ni2-N3 1.946(4), Ni3-C62 1.880(5), Ni3-C32 1.974(6), Ni3-Cl1 2.285(1), Ni3-N1 1.948(4), C2-Ni1-C12 88.3(2), C42-Ni2-C22 87.7(2), C62-Ni3-C32 87.6(2)

3.3 Summary

We have developed a nickel-mediated coupling reaction of 1,2-*o*-carboryne with a variety of alkenes, which gives alkenylcarboranes in moderate to very good isolated yields with excellent regio-and stereoselectivity. This serves a new methodology for the synthesis of alkenylcarboranes. This work also demonstrates that Ni-1,2-*o*-carboryne exhibits different reactivity patterns toward alkynes and alkenes.

The β-H elimination reactions may be suppressed due to steric reasons or intramolecular coordination of the heteroatom leading to the formation of alkylcarboranes after hydrolysis. The thermodynamically stable inserted intermediate offered us an opportunity to investigate its reactivity and to synthesize novel carborane derivatives.

References

1. Hoffmann RW (1967) Dehydrobenzene and cycloalkynes. Academic Press, New York
2. Gilchrist TL (1983) In: Patai S, Rappoport Z (eds) Chemistry of functional groups, supplement C. (Chapter 11) , Wiley, Chichester
3. Buchwald SL, Nielsen RB (1988) Chem Rev 88:1047
4. Hart H (1994) In: Patai S (ed) Chemistry of triple-bonded functional groups, supplement C2. (Chapter 18), Wiley, Chichester
5. Jones WM, Klosin J (1998) Adv Organomet Chem 42:147
6. Ghosh T, Gingrich HL, Kam CK, Mobraaten ECM, Jones M Jr (1991) J Am Chem Soc 113:1313
7. Huang Q, Gingrich HL, Jones M Jr (1991) Inorg Chem 30:3254
8. Cunningham RT, Bian N, Jones M Jr (1994) Inorg Chem 33:4811
9. Ho DM, Cunningham RJ, Brewer JA, Bian N, Jones M Jr (1995) Inorg Chem 34:5274
10. Barnett-Thamattoor L, Zheng G, Ho DM, Jones M Jr, Jackson JE (1996) Inorg Chem 35:7311
11. Atkins JH, Ho DM, Jones M Jr (1996) Tetrahedron Lett 37:7217
12. Lee T, Jeon J, Song KH, Jung I, Baik C, Park K-M, Lee S S, Kang S O, Ko J (2004) Dalton Trans, p 933
13. Jeon J, Kitamura T, Yoo B-W, Kang S O, Ko J (2001) Chem Commun, p 2110
14. Deng L, Chan H-S, Xie Z (2005) J Am Chem Soc 127:13774
15. Ren S, Chan H-S, Xie Z (2009) Organometallics 28:4106
16. Deng L, Chan H-S, Xie Z (2006) J Am Chem Soc 128:7728
17. Bennett M A, Glewis M, Hockless DCR, Wenger E (1997) J Chem Soc, Dalton Trans, p 3105
18. Sayler AA, Beall H, Sieckhaus JF (1973) J Am Chem Soc 95:5790
19. Sicher J (1972) Angew Chem Int Ed 11:200
20. Catellani M, Marmiroli B, Fagnola MC, Acquotti D (1996) J Organomet Chem 507:157
21. Bennett MA, Hockless DCR, Wenger E (1995) For insertion of alkene into Ni-C bond. Organometallics 14:2091
22. Huang D-J, Rayabarapu DK, Li L-P, Sambaiah T, Cheng C-H (2000) Chem Eur J 6:3706
23. Ikeda S-I, Sanuki R, Miyachi H, Miyashita H (2004) J Am Chem Soc 126:10331
24. McDermott JX, White JF, Whitesides GM (1973) J Am Chem Soc 95:4451
25. de Bruin TJM, Magna L, Raybaud P, Toulhoat H (2003) Organometallics 22:3404
26. Tobisch S, Ziegler T (2003) Organometallics 22:5392
27. Huang X, Zhu J, Lin Z (2004) Organometallics 23:4154
28. Xie Z (2006) Coord Chem Rev 250:259
29. Sun Y, Chan H-S, Zhao H, Lin Z, Xie Z (2006) Angew Chem Int Ed 45:5533
30. Sun Y, Chan H-S, Xie Z (2006) Organometallics 25:3447
31. Shen H, Xie Z (2009) Chem Commun, p 2431

Chapter 4
Nickel-Mediated/Catalyzed Three-Component Cycloaddition Reaction of 1,2-*o*-Carboryne/Arynes, Alkenes, and Alkynes

4.1 Introduction

Transition metal-mediated cycloadditions of alkynes and/or alkenes serve as a powerful strategy to construct a wide range of compounds since complexation of the metal center to an olefin or alkyne significantly modifies the reactivity of this moiety [1–4].

1,2-*o*-Carboryne can react with alkenes in ene- and [2+2] reaction manner [5–13]. In contrast, nickel-1,2-*o*-carboryne reacts with alkenes to afford the "Herk type" and "ene-reaction type" cross-coupling products. In the reaction of nickel-1,2-*o*-carboryne with alkenes, when methyl acrylate or 2-vinylpyridine was used as the starting material, only alkylcarboranes were obtained after hydrolysis. The nickelacyclopentane intermediates were isolated in which the donor atom of the olefin can stabilize the intermediates, preventing the β-H elimination. We then studied the reactivity of these intermediates and found that they can react readily with alkynes to give three-component [2+2+2] cycloaddition products.

Multicomponent cross-coupling reactions are a powerful strategy to assemble complex molecules from very simple precursors in a single operation [14]. Arynes, a class of very reactive analogues of alkynes, have recently been reported to undergo metal-catalyzed conversion [15–38]. For examples, the cyclotrimerization of arynes [15–17] and the cocyclization of arynes with alkynes [18–22], allylic halides [23, 24], or activated alkenes [25] can all be catalyzed by palladium. Palladium can also catalyze three-component cross-coupling reactions of arynes, allylic halides [26–31] (allylic epoxides [32], aromatic halides [33]) and alkynylstannanes [26] (boronic acids [27]) to form substituted benzenes, and three-component cyclization of arynes, aryl halides, and alkynes [34, 35] or alkenes [36] to produce phenanthrene derivatives. In contrast, nickel-catalyzed transformations of arynes is much less explored [37, 38].

In view of the analogy between 1,2-*o*-carboryne and its two dimensional relative benzyne, we are also interested in exploring the three-component reaction of benzyne with alkenes and alkynes. In this section we will describe the nickel-mediated

---

Z. Qiu, *Late Transition Metal-Carboryne Complexes*, Springer Theses
DOI: 10.1007/978-3-642-24361-5_4, © Springer-Verlag Berlin Heidelberg 2012

three-component cycloaddition reaction of 1,2-*o*-carboryne with alkenes and alkynes to afford dihydrobenzo-1,2-*o*-carboranes and nickel-catalyzed three-component cycloaddition reaction of arynes with alkenes and alkynes to afford dihydronaphthalenes.

4.2 Results and Discussion

The mono-alkene insertion species **III-9p** or **III-9q** do not show any activity toward olefins such as styrene and 1-hexene (except for excess of activated alkenes such as methyl acrylate and 2-vinylpyridine). But they can react readily with alkynes to give three-component [2+2+2] cycloaddition products. In an initial attempt, the THF solution of **III-9q** or **III-9p** was added with 10 equiv of 3-hexyne and heated at 110 °C for 3 days to afford dihydrobenzo-1,2-*o*-carborane **IV-1a** and **IV-1h** in 92 and 91% isolate yields, respectively (Scheme 4.1). These results show that alkynes are more reactive than alkenes toward these nickelacyclopentane complexes.

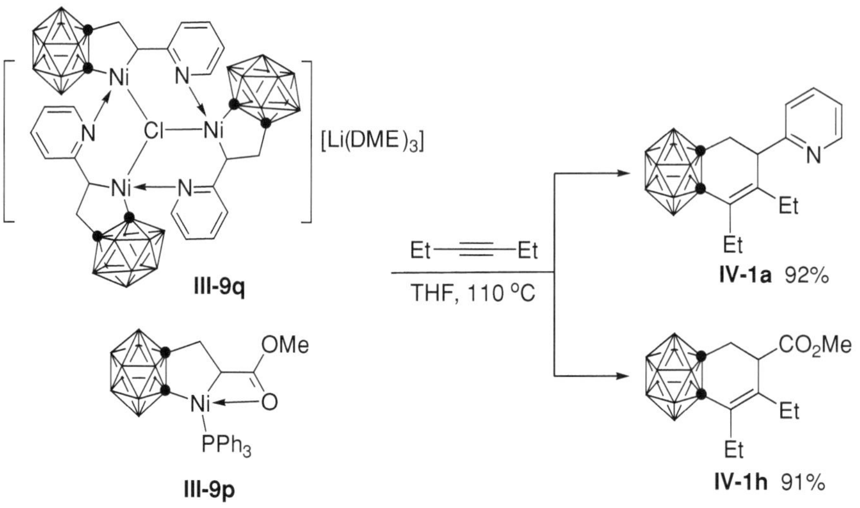

Scheme 4.1 Reactions of nickelacyclopentanes with 3-hexyne

We then examined the reaction in the three-component manner. In a typical procedure, alkene (1.2 equiv) and alkyne (4 equiv) were added to a THF solution of nickel-1,2-*o*-carboryne, prepared in situ by the reaction of $Li_2C_2B_{10}H_{10}$ with $NiCl_2(PPh_3)_2$ [39], and the reaction mixture was heated at 110 °C in a closed vessel. Standard workup procedures afforded the cyclization products in very good chemo- and regio-selectivity (Table 4.1). An excess amount of alkynes were

4.2 Results and Discussion

Table 4.1 Nickel-mediated three-component cycloaddition

Entry	R^1	R^2	R^3	Products	Yield[a]
1	2-Py	Et	Et	**IV-1a**	57
2	2-Py	^{n}Bu	^{n}Bu	**IV-1b**	32
3	2-Py	Me	^{i}Pr	**IV-1c**	34
		^{i}Pr	Me	**IV-1c′**	(1.6:1)
4	2-Py	Me	Ph	**IV-1d**	40
5	2-Py	Me	p-Tolyl	**IV-1e**	35
6	2-Py	Et	Ph	**IV-1f**	39
7	2-Py	^{n}Bu	Ph	**IV-1g**	31
8	2-Py	Ally	Ph	**IV-1h**	36
9	2-Py	Ph	Ph	NR	–
10	2-Py	TMS	TMS	NR	–
11	CO_2Me	Et	Et	**IV-1i**	59
12	CO_2Me	^{n}Pr	^{n}Pr	**IV-1j**	50
13	CO_2Me	^{n}Bu	^{n}Bu	**IV-1k**	48
14	CO_2Me	Ph	Ph	NR	–
15	CO_2Me	TMS	TMS	NR	–

[a] Isolated Yields

necessary as hexasubstituted benzenes were isolated from all reactions, which were generated via Ni-mediated cyclotrimerization of alkynes [1–4]. It is noted that alkynes do not react with nickelacyclopentanes till the reaction temperature reaches ~ 80 °C and the optimal temperature is 110 °C as suggested by GC–MS analyses. On the other hand, activated alkenes can react well with Ni-1,2-o-carboryne in THF at room temperature to give the nickelacycls. Therefore, a separate addition of alkene and alkyne is not necessary for this system.

As shown in Table 4.1, a variety of alkynes are compatible with this nickel-mediated three-component cyclization. Steric factors played an important role in the reactions. Sterically less demanding 3-hexyne offered the highest yield (entries 1 and 11). No reaction proceeded for diphenylacetylene (entries 9 and 14) and bis(trimethylsilyl)acetylene (entries 10 and 15). 4-Methyl-2-pentyne offered two regio-isomers in a molar ratio of 1.6:1 (entry 3). Other unsymmetrical alkynes gave only one isomer of **IV-1** due to the electronic effects as phenyl can be viewed as electron-withdrawing group (entries 4–8) [40, 41]. In the case of $CH_2=CHCH_2C\equiv CC_6H_5$, no C=C insertion product was observed (entry 8). It is noteworthy that terminal alkynes quenched the reaction intermediates to afford **III-9q,r**, and nitriles, isonitriles, or carbodiimides did not yield any insertion products.

Compounds **IV-1** were fully characterized by ^{1}H, ^{13}C, and ^{11}B NMR as well as high-resolution mass spectrometry. The regioisomers of **IV-1c** were assigned using NOESY analyses (Chart 4.1).

Chart 4.1 Assignment of the regioisomer **IV-1c**

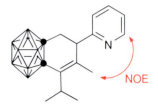

In the ^{1}H NMR spectra (CDCl$_3$) of **IV-1a–1h**, which generated from the reaction of Ni-1,2-o-carboryne, 2-vinylpyridine, and alkynes, two doublet of doublets at ~2.8 ppm with ^2J = 14.8 Hz, ^3J = 7.2 Hz and ~3.0 ppm with ^2J = 14.8 Hz, ^3J = 10.8 Hz assignable to the carborane cage connected CH$_2$ unit and one doublet of doublet at ~3.9 ppm with ^3J = 7.2 and 10.8 Hz corresponding to the CH proton, were observed. In case of methyl acrylate insertion products **IV-1i–1k**, the CH signal was shifted upfield to ~3.3 ppm as a multiplet. The two CH$_2$ signals moved to ~2.5 ppm with ^2J = 14.7 Hz, ^3J = 7.3 Hz and ~3.2 pp with ^2J = 14.7 Hz, ^3J = 6.0 Hz. Their ^{13}C NMR spectra were consistent with the ^{1}H NMR results. The ^{13}C NMR spectra showed the signals of CH and CH$_2$ unit at 47–41 and 37–32 ppm, respectively. The ^{11}B NMR spectra exhibited a 1:1:8 splitting pattern for **IV-1a–1h** and a 1:1:1:1:6 splitting pattern for **IV-1i–1k**.

The solid-state structures of **IV-1d** and **IV-1i** were further confirmed by single-crystal X-ray analyses. In the molecular structure of **IV-1d** (Fig. 4.1), the bond distances and angles indicate that C(11) and C(12) are sp^3-carbons whereas C(18) and C(20) are sp^2-carbons. There are two crystallographically independent molecules in the unit cell of **IV-1i**. Figure 4.2 shows the representative structure. The bond distances and angles are very close to those observed in **IV-1d**.

Scheme 4.2 shows the plausible mechanism for the formation of [2+2+2] cycloaddition products. The trinuclear Ni complex in **III-9r** may be dissociated into mononuclear Ni complex during the reaction. Accordingly, the formation of products **IV-1** can be rationalized by the sequential insertion of alkene and alkyne into the Ni–C bond. The insertion of alkene affords the nickelacyclopentane **III-9**. Subsequent insertion of alkyne into the nickel–C(alkyl) bond gives the seven-membered intermediate **IV-A** [42–46]. Reductive elimination yields the final products **IV-1**.

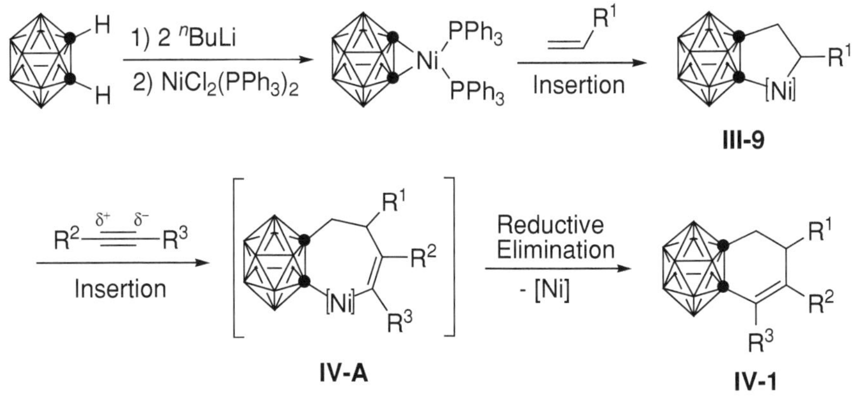

4.2 Results and Discussion

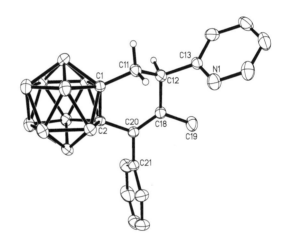

Scheme 4.2 Proposed mechanism for three-component cycloaddition

Fig. 4.1 Molecular structure of **IV-1d**. Selected bond lengths (Å) and angles (deg): C1–C2 1.640(2), C1–C11 1.519 (2), C11–C12 1.536(2), C12–C18 1.531(2), C18–C20 1.340(2), C2–C20 1.503(2), C2–C1–C11 115.8(1), C1–C11–C12 114.4(1), C11–C12–C18 112.8(1), C12–C18–C20 122.6(1), C18–C20–C2 122.2(1), C20–C2–C1 115.9(1)

Fig. 4.2 Molecular structure of **IV-1i**. Selected bond lengths (Å) and angles (deg): C1–C2 1.650(3), C1–C11 1.507 (4), C11–C12 1.516(4), C12–C13 1.532(4), C13–C14 1.335(4), C2–C14 1.493(4), C2–C1–C11 115.6(2), C1–C11–C12 113.5(2), C11–C12–C13 114.7(2), C12–C13–C24 122.5(2), C13–C14–C2 121.6(2), C14–C2–C1 116.5(2)

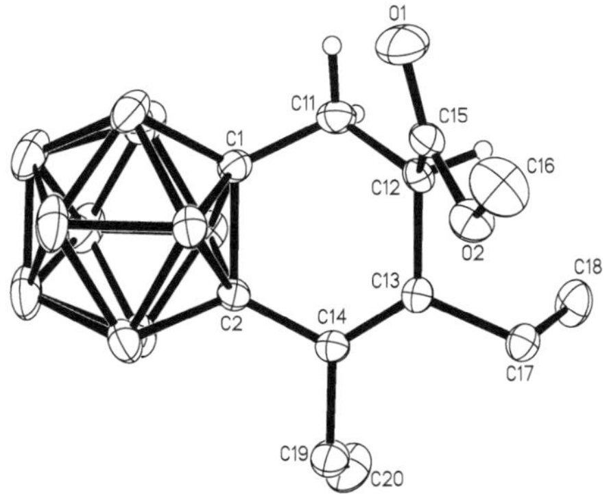

We then extended our research to include arynes and found that nickel can efficiently catalyze three-component [2+2+2] cyclization of arynes, alkenes, and alkynes to afford a series of substituted dihydronaphthalenes that cannot be prepared from readily available starting materials [47–55].

In an initial attempt, a CH_3CN solution of benzyne precursor **IV-2a** [1 equiv, 2-(trimethylsilyl)phenyltriflate], methyl acrylate **IV-3a** (2 equiv), and diphenyl acetylene **IV-4a** (1.2 equiv) in the presence of Ni(cod)$_2$ (10 mol %) and CsF (3 equiv) was stirred at room temperature for 5 h to give the cyclization product **IV-5a** in 72% isolated yield (Scheme 4.3 and Table 4.2, entry 6).

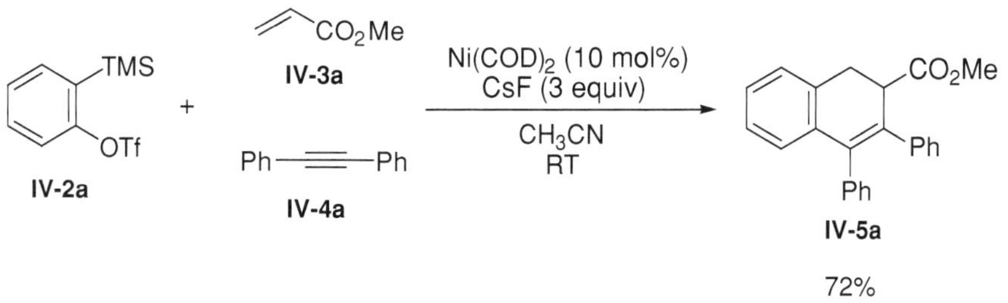

Scheme 4.3 Three-component reaction of benzyne, methyl acrylate, and diphenylacetylene

Subsequent work focused on optimization of this reaction (Table 4.2). Changing the ligand from cod to PPh$_3$ or adding PPh$_3$ to Ni(cod)$_2$ led to a big drop in the isolation of **IV-5a** from 72 to 50% yield (entries 6 and 9). Addition of bidentate ligand dppe further decreased the isolated yield of **IV-5a** to 21% (entry 10). No detectable **IV-5a** was observed when $NiCl_2\left(PBu_3^n\right)_2$/Zn or NiCl$_2$(dppp)/ Zn was used as catalyst (Table 4.2, entries 3 and 5). In contrast, palladium complexes such as Pd(dba)$_2$, PdCl$_2$(PPh$_3$)$_2$/Zn, and Pd(PPh$_3$)$_4$ did not mediate three-component benzyne–alkene–alkyne cyclization, rather they catalyzed two-component benzyne–alkene–benzyne cycloaddition and cross-coupling [25] to afford 9,10-dihydrophenanthrene **IV-6a** and methyl 3-(1,1′-biphenyl-2-yl)-2-propenate **IV-7a** (entries 11–13).

These results showed that (1) both metal and ligand had significant effects on the reactions; (2) activated alkene is more reactive than alkyne, otherwise two-component benzyne–alkyne–alkyne cycloaddition products should be observed; and (3) Ni(cod)$_2$ exhibited the highest catalytic activity in three-component [2+2+2] cyclization. The same results were observed when the catalyst loading was decreased from 10 to 5 mol % (Table 4.2, entry 8) or the reaction temperature was increased from room temperature 20 to 50 °C (Table 4.2, entry 7).

The scope and limitation of this Ni-catalyzed cyclization process were then examined using various alkenes and aryne precursors. The results were summarized in Table 4.3. Acrylates **IV-3a,b,c** gave very high isolated yields (72–76%) of the corresponding cocyclization products **IV-5a,b,c** (entries 1–3). Methyl vinyl ketone **IV-4d** and acrylonitrile **IV-5e** offered very low yields of the desired aryne–alkene–alkyne cocyclization product of **IV-5d** (3%) and **IV-5e** (15%) (entries 4

4.2 Results and Discussion

Table 4.2 Optimization of three-component cycloaddition reaction[a]

Entry	Catalyst	Loading (mol %)	Yield of **IV-5a** (%)[b] (**IV-5a:IV-6a:IV-7a**)[c]
1	Ni(PPh$_3$)$_4$	10	50 (55: <2:43)
2	NiCl$_2$(PPh$_3$)$_2$/Zn (1:3)	10	52 (67:8:25)
3	NiCl$_2$(PBu$_3^n$)$_2$/Zn (1:3)	10	0 (<2:51:47)
4	NiCl$_2$(dppe)/Zn (1:3)	10	11 (17:32:51)
5	NiCl$_2$(dppp)/Zn (1:3)	10	0 (<1:27:72)
6	Ni(cod)$_2$	10	72 (90: <5: <5)
7	Ni(cod)$_2$	10	73[d] (90: <5: <5)
8	Ni(cod)$_2$	5	72 (90: <5: <5)
9	Ni(cod)$_2$/PPh$_3$ (1:2)	10	51 (55: <2:43)
10	Ni(cod)$_2$/dppe (1:1)	10	21 (23:16:61)
11	Pd(dba)$_2$	10	0 (<2:37:61)
12	PdCl$_2$(PPh$_3$)$_2$/Zn (1:3)	10	0 (<2:52:46)
13	Pd(PPh$_3$)$_4$	10	0 (<1:88:11)

[a] Condition: **IV-2a** (0.3 mmol), **IV-3a** (0.6 mmol), **IV-4a** (0.36 mmol), and CsF (0.9 mmol) in CH$_3$CN (1 mL) at r.t. for 5 h
[b] Isolated yields of **IV-5a**
[c] Ratio determined by ^{1}H NMR spectroscopy on the crude product mixture
[d] The reaction was carried out at 50 °C

and 5). In these reactions, the major products were aryne–alkene–aryne cyclization species. If unactivated alkenes were used, no desired products **IV-5** were detected. Functionalized aryne precursors with electron-donating groups (**IV-2c,d,e**) were less effective, producing dihydronaphthalene derivatives **IV-5f,g,h/h′** in moderate yields (entries 7–9). The electron-poor benzyne precursor **IV-2b** afforded an inseparable complex mixture (entry 6).

A variety of alkynes were compatible with this nickel-catalyzed cocyclization reaction and gave the desired products **IV-5** in very good yields (Table 4.4). An excellent regioselectivity was observed for all unsymmetrical alkynes because of the polarity of these molecules (entries 1–8). It is noteworthy that no C=C or

62   4 Nickel-Mediated/Catalyzed Three-Component Cycloaddition Reaction

Table 4.3 Nickel-catalyzed three-component cycloaddition of arynes with activated alkenes and diphenylacetylene

Entry	IV-2	IV-3	IV-5	Yield (%)[b]
1	IV-2a	IV-3a (CO_2Me)	IV-5a (CO_2Me)	76
2	IV-2a	IV-3b ($CO_2{}^nBu$)	IV-5b ($CO_2{}^nBu$)	72
3	IV-2a	IV-3c ($CO_2{}^tBu$)	IV-5c ($CO_2{}^tBu$)	74
4	IV-2a	IV-3d	IV-5d	3
5	IV-2a	IV-3e (CN)	IV-5e (CN)	15
6	IV-2b	IV-3a	–	–

(continued)

4.2 Results and Discussion

Table 4.3 (continued)

Entry	IV-2	IV-3	IV-5	Yield (%)[b]
7	IV-2c	IV-3a	IV-5f	29[c]
8	IV-2d	IV-3a	IV-5g	46[c]
9	IV-2e	IV-3a	IV-5h / IV-5h′	57[c] (IV-5h:IV-5h′ = 1.3:1)[d]

[a] Condition: **IV-2** (0.3 mmom), **IV-3** (0.6 mmol), **IV-4a** (0.6 mmol), CsF (0.9 mmol) in CH$_3$CN (1 mL) at r.t. for 5 h
[b] Isolated yields
[c] The reaction was carried out at r.t. overnight
[d] The ratio was estimated by ^{1}H NMR

$C \equiv N$ insertion product was observed when **IV-4f** or **IV-4g** was used as the starting material (entries 5 and 6). In case of methyl 2-butynoate **IV-4i**, the low yield was due to the trimerization of **IV-4i** catalyzed by Ni(0) (entry 8) [1–4, 56]. It's known that an alkyne with electron-withdrawing substituents is more reactive than one with electron-donating groups in Ni(0)- or Pd(0)-catalyzed [2+2+2] co-cyclization [21, 57, 58], which is consistent with the result of entries 9–11. Dialkylacetylene **IV-4j–l** (2 equiv) afforded **IV-5q–s/s′** in 22–44% yields. When 3 equiv of alkyne was used, the yields can be raised to 47–63%. And two regioisomers **IV-5s** and **IV-5s′** were obtained in 2:1 molar ratio in the reaction of 4-methyl-2-pentyne **IV-4l**.

Compounds **IV-5** were fully characterized by ^{1}H and ^{13}C NMR as well as high-resolution mass spectrometry. In the ^{1}H NMR spectrum of **IV-5**, three doublet of

Table 4.4 Nickel-catalyzed three-component cycloaddition of benzyne with methyl acrylate and alkynes[a]

Entry	IV-4	IV-5	Yield (%)[b]
1	Me━━━Ph **IV-4b**	**IV-5i**	71
2	Et━━━Ph **IV-4c**	**IV-5j**	78
3	Bu^n━━━Ph **IV-4d**	**IV-5k**	71
4	MeO⌒━━━Ph **IV-4e**	**IV-5l**	75
5	⌒━━━Ph **IV-4f**	**IV-5m**	68
6	NC⌒⌒━━━Ph **IV-4g**	**IV-5n**	32

(continued)

4.2 Results and Discussion

Table 4.4 (continued)

Entry	IV-4	IV-5	Yield (%)[b]
7	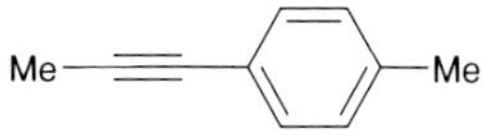 IV-4h	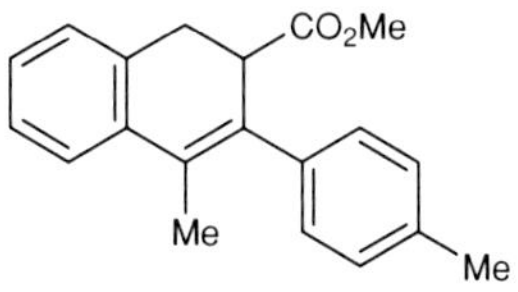 IV-5o	66
8	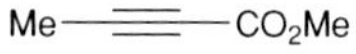 IV-4i	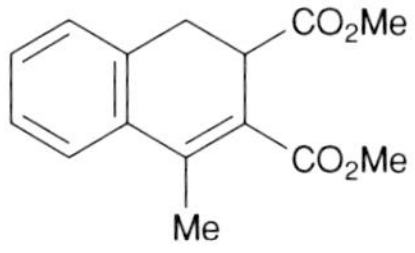 IV-5p	19
9	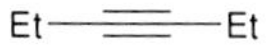 IV-4j	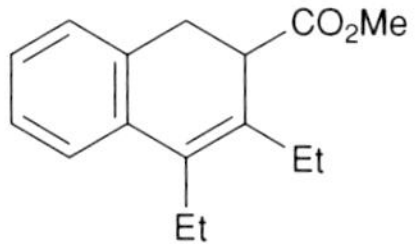 IV-5q	31(63)[c]
10	 IV-4k	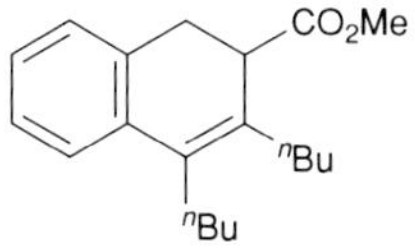 IV-5r	22(47)[c]
11	 IV-4l	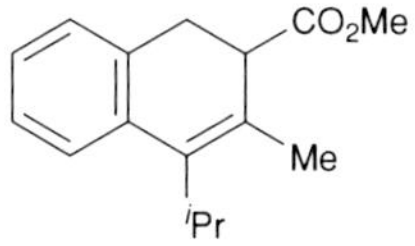 IV-5s 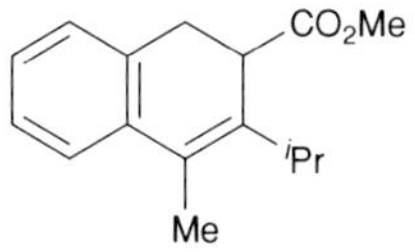 / IV-5s'	44(51)[c] (**IV-5s:IV-5s'** = 2:1)[d]

[a] Condition: **IV-2a** (0.3 mmol), **IV-3a** (0.6 mmol), **IV-4** (0.6 mmol), and CsF (0.9 mmol) in CH$_3$CN (1 mL) at r.t. for 5 h

[b] Isolated yields

[c] 3 mmol of Alkyne was used

[d] The ratio was estimated by [1] H NMR

doublets or one triplet and two doublet of doublets can be observed at 3–4 ppm corresponding to the CH and CH_2 protons. Their ^{13}C NMR spectra were also consistent with the results observed with **IV-1**, which showed the signals of CH and CH_2 unit at ~ 46 and ~ 32 ppm, respectively. The relative regiochemical assignments of **IV-5h** and **IV-5h$'$** were determined using HH COSY analyses and the diagnostic correlation is shown in Chart 4.2.

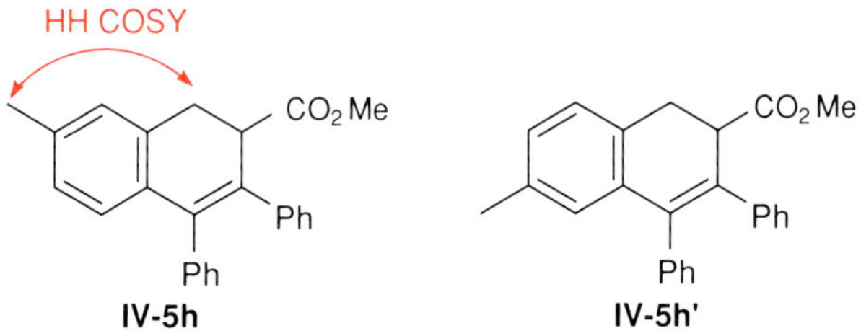

Chart 4.2 Assignment of the regioisomers of **IV-5h** and **IV-5h$'$**

The relative regiochemical assignments of 1,2-dihydronaphthalene **IV-5** were determined using NOESY analyses and the diagnostic correlations are shown in Chart 4.3.

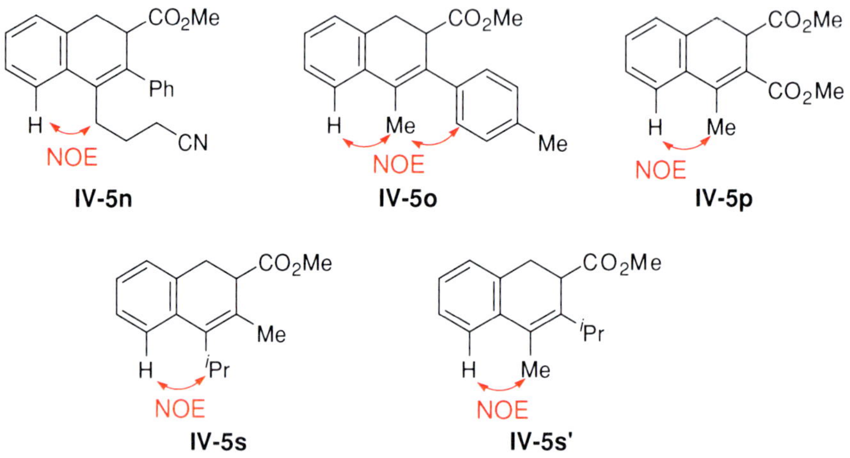

Chart 4.3 Assignment of regioisomers

The molecular structures of **IV-5e** and **IV-5l** were further confirmed by single-crystal X-ray analyses (Figs. 4.3, 4.4).

4.2 Results and Discussion

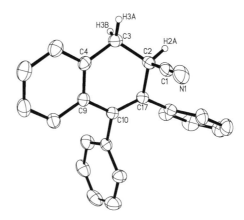

Fig. 4.3 Molecular structure of **IV-5e**. Selected bond lengths (Å) and angles (deg): C4–C9 1.401(3), C4–C3 1.500 (3), C3–C2 1.524(3), C2–C17 1.537(2), C17–C10 1.342(2), C10–C9 1.476(3), C9–C4–C3 118.2(2), C4–C3–C2 112.1(2), C3–C2–C17 111.5(2), C2–C17–C10 119.3(2), C17–C10–C9 121.3(2), C10–C9–C4 119.7(2)

A plausible mechanism for the nickel-catalyzed three-component cocyclization is shown in Scheme 4.4. The catalysis is likely initiated by oxidative coupling of benzyne and alkene on Ni(0) to form a nickelacycle **IV-B,** which is probably stabilized by an intramolecular coordination of the heteroatom. Subsequent insertion of alkyne into the nickel–C(aryl) bond gives the seven-membered intermediate **IV-C** [40, 41]. The regioselectivity observed in the reactions can be rationalized by the polarity of alkynes [41]. Reductive elimination of **IV-B** yields the final product **IV-5** and regenerates the catalyst.

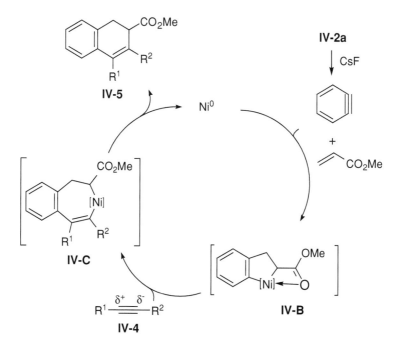

Scheme 4.4 Proposed mechanism of nickel-catalyzed [2+2+2] cyclization reaction

Fig. 4.4 Molecular structure of **IV-5l**. Selected bond lengths (Å) and angles (deg): C8–C3 1.404(3), C3–C2 1.500 (3), C2–C1 1.530(3), C1–C10 1.517(3), C10–C9 1.347(2), C9–C8 1.490(2), C8–C3–C2 118.5(2), C3–C2–C1 112.7(2), C2–C1–C10 112.5(2), C1–C10–C9 120.8(2), C10–C9–C8 120.8(2), C9–C8–C3 119.5(2)

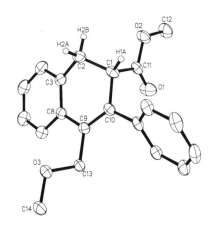

4.3 Summary

We have developed a novel nickel-mediated three-component assembling reaction of 1,2-*o*-carboryne with alkenes and alkynes and a novel nickel-catalyzed three-component [2+2+2] carboannulation of arynes, activated alkenes, and alkynes. This work offers an exceptionally efficient route to the synthesis of di-hydrobenzocarborane and 1,2-dihydronaphthalenes derivatives from readily available starting materials.

References

1. Schore NE (1988) Chem Rev 88:1081
2. Trost BM (1991) Science 254:1471
3. Lautens M, Klute W, Tam W (1996) Chem Rev 96:49
4. Saito S, Yamamoto Y (2000) Chem Rev 100:2901
5. Gingrich HL, Ghosh T, Huang Q, Jones M Jr (1990) J Am Chem Soc 112:4082
6. Ghosh T, Gingrich HL, Kam CK, Mobraaten ECM, Jones M Jr (1991) J Am Chem Soc 113:1313
7. Huang Q, Gingrich HL, Jones M Jr (1991) Inorg Chem 30:3254
8. Cunningham RT, Bian N, Jones M Jr (1994) Inorg Chem 33:4811
9. Ho DM, Cunningham RJ, Brewer JA, Bian N, Jones M Jr (1995) Inorg Chem 34:52[74]
10. Barnett-Thamattoor L, Zheng G, Ho DM, Jones M Jr, Jackson JE (1996) Inorg Chem 35:7311
11. Atkins JH, Ho DM, Jones M Jr (1996) Tetrahedron Lett 37:7217
12. Lee T, Jeon J, Song KH, Jung I, Baik C, Park K-M, Lee SS, Kang SO, Ko J (2004) Dalton Trans 933
13. Jeon J, Kitamura T, Yoo B-W, Kang SO, Ko (2001) J Chem Commun 2110
14. Zhu J, Bienaymé H (eds) (2005) Multicomponent reactions. Wiley-VCH, Weinheim, Germany
15. Peña D, Escudero S, Pérez D, Guitián E, Castedo L (1998) Angew Chem Int Ed 37:2659
16. Peña D, Pérez D, Guitián E, Castedo L (1999) Org Lett 1:1555
17. Romero C, Peña D, Pérez D, Guitián E (2006) Chem Eur J 12:5677
18. Peña D, Pérez D, Guitián E, Castedo LJ (1999) Am Chem Soc 121:5827

References

19. Radhakrishnan KV, Yoshikawa E, Yamamoto Y (1999) Tetrahedron Lett 40:7533
20. Peña D, Pérez D, Guitián E, Castedo LJ (2000) Org Chem 65:6944
21. Sato Y, Tamura T, Mori M (2004) Angew Chem Int Ed 43:2436
22. Romero C, Peña D, Pérez D, Guitián EJ (2008) Org Chem 73:7996
23. Yoshikawa E, Radhakrishnan KV, Yamamoto Y (2000) J Am Chem Soc 122:7280
24. Yoshikawa E, Yamamoto Y (2000) Angew Chem Int Ed 39:173
25. Quintana I, Boersma AJ, Peña D, Pérez D, Guitián E (2006) Org Lett 8:3347
26. Jeganmohan M, Cheng C-H (2004) Org Lett 6:2821
27. Jayanth TT, Jeganmohan M, Cheng C-H (2005) Org Lett 7:2921
28. Henderson JL, Edwards AS, Greaney MF (2006) J Am Chem Soc 128:7426
29. Bhuvaneswari S, Jeganmohan M, Yang M-C, Cheng C-H (2008) Chem Commun 2158
30. Xie C, Liu L, Zhang Y, Xu P (2008) Org Lett 10:2393
31. Bhuvaneswari S, Jeganmohan M, Cheng C-H (2008) Chem Commun 5013
32. Jeganmohan M, Bhuvaneswari S, Cheng C-H (2009) Angew Chem Int Ed 48:391
33. Henderson JL, Edwards AS, Greaney MF (2007) Org Lett 9:5589
34. Liu Z, Zhang X, Larock RC (2005) J Am Chem Soc 127:15716
35. Liu Z, Larock RC (2007) Angew Chem Int Ed 46:2535
36. Bhuvaneswari S, Jeganmohan M, Cheng C-H (2006) Org Lett 8:5581
37. Hsieh J-C, Cheng C-H (2005) Chem Commun 2459
38. Jayanth TT, Cheng C-H (2007) Angew Chem Int Ed 46:5921
39. Sayler AA, Beall H, Sieckhaus JF (1973) J Am Chem Soc 95:5790
40. Bennett MA, Glewis M, Hockless DCR, Wenger E (1997) J Chem Soc, Dalton Trans 3105
41. Bennett MA, Macgregor SA, Wenger E (2001) Helv Chim Acta 84:3084
42. Deng L, Chan H-S, Xie Z (2006) J Am Chem Soc 128:7728
43. Xie Z (2006) Coord Chem Rev 250:259
44. Sun Y, Chan H-S, Zhao H, Lin Z, Xie Z (2006) Angew Chem Int Ed 45:5533
45. Sun Y, Chan H-S, Xie Z (2006) Organometallics 25:3447
46. Shen H, Xie Z (2009) Chem Commun 2431
47. Wennerberg J, Ek F, Hansson A, Frejd T (1999) J Org Chem 64:54
48. Hamura T, Miyamoto M, Imura K, Matsumoto T, Suzuki K (2002) Org Lett 4:1675
49. Inoue H, Chatanl N, Murai S (2002) J Org Chem 67:1414
50. Asao N, Kasahara T, Yamamoto Y (2003) Angew Chem Int Ed 42:3504
51. Zhou H, Huang X, Chen W (2004) J Org Chem 69:5471
52. Lautens M, Schmid GA, Chau A (2002) J Org Chem 67:8043
53. Rayabarapu DK, Chiou C-F, Cheng C-H (2002) Org Lett 4:1679
54. Wu M-S, Rayabarapu DK, Cheng C-H (2004) J Org Chem 69:8407
55. Wu M-S, Jeganmohan M, Cheng C-H (2005) J Org Chem 70:9545 (and references therein)
56. Jhingan AK, Maier WF (1987) J Org Chem 52:1161
57. Brown LD, Itoh K, Suzuki H, Hirai K, Ibers JA (1978) J Am Chem Soc 100:8232
58. Stephan C, Munz C, Dieck H (1993) J Organomet Chem 452:223

Chapter 5
Nickel-Catalyzed [2+2+2] Cycloaddition of 1,2-o-Carboryne with Alkynes

5.1 Introduction

Reactivity studies showed that 1,2-o-carboryne can react with alkenes, dienes, and alkynes in [2+2], [2+4] cycloaddition and ene-reaction patterns, [1–9] similar to that of benzyne. The carboryne reactions are usually complicated and not in a controlled manner. On the other hand, nickel-1,2-o-carboryne complex $(\eta^2\text{-}C_2B_{10}H_{10})Ni(PPh_3)_2$ [10] can undergo regioselective [2+2+2] cycloaddition reactions with 2 equiv of alkynes to afford benzocarboranes [11]. This reaction requires a stoichiometric amount of Ni reagent. Transition metal-catalyzed cocyclization of π-component molecules has received much attention because of its highly atom-economical nature [12–15]. In view of the analogy between metal-benzyne [16–20] and metal-1,2-o-carboryne complexes and the metal-catalyzed reactions of benzyne with alkenes and alkynes [21–44], we wondered if a catalytic version of nickel-mediated reactions of 1,2-o-carboryne could be developed.

In the stoichiometric reactions, high temperature was necessary for the insertion of alkynes into the Ni–C_{cage} bond in Ni-1,2-o-carboryne and the Ni(0) species was the end metal complex [45]. On the other hand, 1-bromo-2-lithiocarborane was reported as a precursor of 1,2-o-carboryne [1–7]. It is rational to assume that 1-bromo-2-lithiocarborane may undergo oxidative addition with Ni(0) to give, after elimination of LiBr, the desired Ni-1,2-o-carboryne complex to construct a catalytic cycle. In this section we will report the nickel-catalyzed [2+2+2] cycloaddition of 1,2-o-carboryne with 2 equiv of alkynes to afford benzocarboranes.

5.2 Results and Discussion

1-Iodo-2-lithiocarborane was chosen as the precursor to realize the catalytic cycle because it is more efficient than 1-bromo-2-lithiocarborane (vide infra) and iodine is easy to handle with. 1-Iodo-2-lithiocarborane, conveniently prepared in situ

---

Z. Qiu, *Late Transition Metal-Carboryne Complexes*, Springer Theses
DOI: 10.1007/978-3-642-24361-5_5, © Springer-Verlag Berlin Heidelberg 2012

Table 5.1 Optimization of reaction conditions[a]

V-1a

Entry	Catalyst[b]	Loading mol (%)	Time (h)	T (°C)	Yield (%)[c]
1	Ni(cod)$_2$	20	2	110	49
2	Ni(cod)$_2$/4PPh$_3$	20	2	110	33
3	Ni(PPh$_3$)$_4$	20	2	110	37
4	NiCl$_2$(PMe$_3$)$_2$	20	2	110	17
5	NiCl$_2$(P^{n}Bu$_3$)	20	2	110	57
6	NiCl$_2$(PPh$_3$)$_2$	20	2	110	65
7	NiCl$_2$(PPh$_3$)$_2$	10	2	110	31
8	NiCl$_2$(PPh$_3$)$_2$	20	4	110	63
9	NiCl$_2$(PPh$_3$)$_2$	20	4	90	60
10	NiCl$_2$(dppe)	20	2	110	29
11	NiCl$_2$(dppp)	20	2	110	22
12	NiI$_2$(Me$_2$Im)$_2$	20	2	110	16
13	Pd(PPh$_3$)$_4$	20	2	110	1
14	PdCl$_2$(PPh$_3$)$_2$	20	2	110	1
15	FeCl$_2$/2PPh$_3$	20	2	110	-
16	CoCl$_2$(PPh$_3$)$_2$	20	2	110	-

[a] Condition: (1) carborane (0.5 mmol), n-BuLi (1.0 mmol), in toluene at room temperature for 1 h, (2) I$_2$ (0.5 mmol), at room temperature for 0.5 h, (3) catalyst, 3-hexyne (2 mmol)
[b] cod = cyclooctadiene, dppe = 1,2-bis(diphenylphosphino)ethane, dppp = 1,3-bis(diphenylphosphino)propane, Me$_2$Im = 1,3-dimethylimidazol-2-ylidene
[c] Isolated yields

from the reaction of dilithiocarborane with 1 equiv of iodine in toluene at room temperature, was thermally stable at room temperature. Heating a benzene solution of 1-iodo-2-lithiocarborane overnight afforded a [4+2] cycloaddition product 1,2-(2,5-cyclohexadiene-1,4-diyl)-o-carborane in 25% isolated yield, which is much higher than the 8% yield from 1-bromo-2-lithiocarborane precursor [7]. This result suggests that 1-iodo-2-lithiocarborane is a more efficient precursor than the bromo one.

We then examined the catalytic activity of various metal complexes in the reaction of 1-iodo-2-lithiocarborane with an excess amount of 3-hexyne in toluene at 110 °C for 2 h. The results were summarized in Table 5.1. The Ni(0) complexes were all catalytically active with Ni(cod)$_2$ being the most active one, giving the desired [2+2+2] cycloaddition product **V-1a** in 33–49% isolated yields (entries 1–3). Addition of PPh$_3$ led to a big drop in the yield of **V-1a** from 49 to 33%, probably suggesting that free PPh$_3$ and alkyne compete the coordination site of the Ni atom. The Ni(II) salts were also active. Their activities depended largely on the

5.2 Results and Discussion

ligands (entries 4–12). NiCl$_2$(PPh$_3$)$_2$ was found to be the best catalyst, producing **V-1a** in 65% isolated yield, suggesting that the in situ generated Ni(0) species is more active than Ni(cod)$_2$ (vide infra) (entry 6). Lower catalyst loading (10 mol%) resulted in a significant decrease of the yield from 65 to 31% (entry 7). Extension of reaction time from 2 to 4 h did not affect the yield of **V-1a** (entry 8).

Temperature was crucial to the reaction. Compound **V-1a** was not observed if the reaction temperatures were <60 °C. The reaction proceeded well at 90 °C, but needed a longer time to completion (entry 9). It was noted that in addition to **V-1a** other products were o-carborane with small amounts of 1,2-o-carboryne–alkyne ene-reaction product **V-3** (Scheme 5.2) and 1,2-o-carboryne–toluene [2+4] cycloaddition reaction products **V-2** (Scheme 5.2) in the above reactions. In sharp contrast, palladium complexes such as PdCl$_2$(PPh$_3$)$_2$ and Pd(PPh$_3$)$_4$ showed almost no activity (entries 13 and 14). FeCl$_2$/2PPh$_3$ and CoCl$_2$(PPh$_3$)$_2$ were inactive (entries 15 and 16).

To further investigate the reaction, we examined the following control experiments. The toluene solution of the in situ prepared 1-iodo-2-lithiocarborane was heated at 110 °C for 2 h. [4+2] cycloaddition products **V-2a** and **V-2b** were obtained as inseparable mixture in 38% isolated yield (Scheme 5.1).

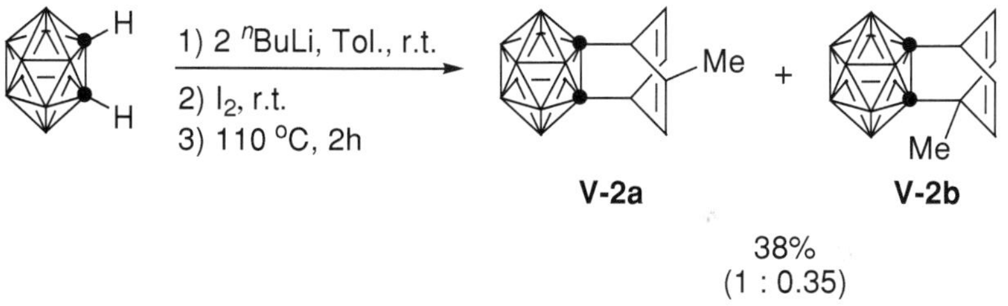

Scheme 5.1 Reaction of 1,2-o-carboryne precursor with toluene

When 4 equiv of 3-hexyne was used, the ene-reaction product **V-3** was obtained as the major product in 36% yield after heating. [4+2] cycloaddition product was isolated in 17% yield, and carborane was recovered in 27% yield (Scheme 5.2).

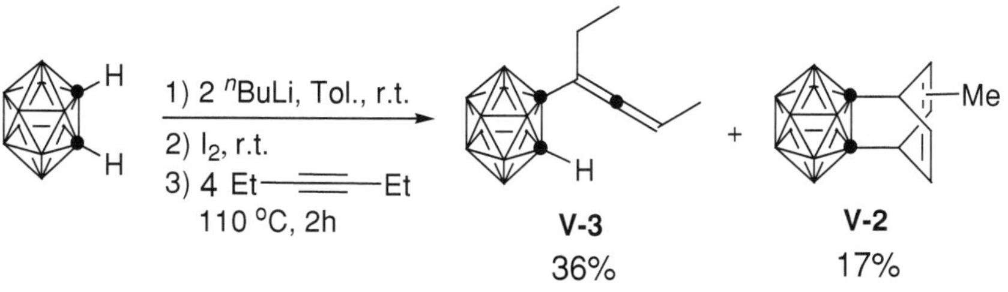

Scheme 5.2 Reaction of 1,2-o-carboryne precursor with 3-hexyne in toluene

Table 5.2 Nickel-catalyzed cycloaddition of 1,2-*o*-carborynes with alkynes

Entry	R^1	R^2/R^3	V-5	Product	Yield (%)[a,b]
1	H	Et/Et	**V-5a**	**V-1a**	65 (67)
2	3-Cl	Et/Et	**V-5a**	**V-1b**	31
3	3-Ph	Et/Et	**V-5a**	**V-1c**	38
4	H	$^nPr/^nPr$	**V-5b**	**V-1d**	59 (65)
5	H	$^nBu/^nBu$	**V-5c**	**V-1e**	54 (65)
6	H	Ph/Ph	**V-5d**	**V-1f**	28 (33)
7	H	CH_2OMe/CH_2OMe	**V-5e**	**V-1g**	13
8	H	Me/^iPr	**V-5f**	**V-1h + V-1'h**	44 (**V-1g/V-1'g** = 70/30)[c]
9	H	Me/Ph	**V-5 g**	**V-1i**	50 (54)
10	H	Me/*p*-Tolly	**V-5h**	**V-1j**	39
11	H	Me/p-CF$_3$–C$_6$H$_4$	**V-5i**	**V-1k**	49
12	H	Et/Ph	**V-5j**	**V-1l**	49
13	H	$^nBu/Ph$	**V-5k**	**V-1m**	43
14	H	C≡CPh/Ph	**V-5l**	**V-1n**	51
15	H	CH_2OMe/Ph	**V-5m**	**V-1o/V-1'o**	24/2
16	H	CH_2NMe_2/Ph	**V-5n**	–	–
17	H	CO_2Me/Me	**V-5o**	–	–

[a] Isolated yields
[b] Yields in parentheses are corresponding to those of stoichiometric reactions of Ni-1,2-*o*-carboryne with 2 equiv of alkynes, reported in Ref. [45]
[c] Molar ratio was determined by ^{1}H NMR spectroscopy on the crude product mixture

5.2 Results and Discussion

This result implies that alkynes are more efficient reagents than arenes in the reaction with 1,2-*o*-carboryne.

When 1-iodo-2-lithiocarborane toluene solution was heated in the presence of 20 mol% NiCl$_2$(PPh$_3$)$_2$, only 15 mol% of 1-iodocarborane was recovered along with the isolation of carborane (53%) (Scheme 5.3). The transition-metal species can largely change the reaction pathway and prohibit the [4+2] cycloaddition reaction of 1,2-*o*-carboryne with toluene.

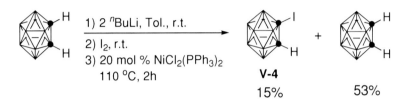

Scheme 5.3 Heating of 1,2-*o*-carboryne precursor in toluene in the presence of NiCl$_2$(PPh$_3$)$_2$

We then expanded the substrates scope of the catalytic cycloaddition reaction to include various carboranes and alkynes using the above optimal reaction condition (Table 5.1, entry 6). The results were compiled in Table 5.2. The isolated yields of **V-1** were very comparable with those of stoichiometric reactions of Ni-1,2-*o*-carboryne with alkynes (entries 1, 4–6, and 9) [45]. Steric factors played an important role in the reactions. Sterically less demanding 3-hexyne offered the highest yield (entry 1). Carboranes with 3-chloro and 3-phenyl resulted in a big decrease in the isolated yields of **V-1b,c** from 65 to 31–38% (entries 2 and 3). 4-Methyl-2-pentyne **V-5f** offered two inseparable regio-isomers **V-1 h/V-1′h** in a molar ratio of 7:3 (entry 8). However, an excellent regioselectivity was observed for unsymmetrical arylalkynes due to the electronic effects as phenyl can be viewed as electron-withdrawing group (entries 9–14) [46, 47]. For alkynes bearing ether groups **V-5e** and **V-5m**, the products were formed in low yields, probably due to the coordination of oxygen atom occupying the vacant site of the Ni atom (entries 7 and 15). Such interactions may also alter the regioselectivity of the alkyne insertion and stabilize the inserted product, which leads to the formation of **V-1′o** and a small amount of mono-alkyne insertion products after hydrolysis (vide infra) (entry 15). Alkynes bearing amido group or carbonyl group such as **V-5n** and **V-5o** were incompatible with this reaction because they can react with the carboryne precursor 1-iodo-2-lithiocarborane (entries 16 and 17). For methyl 2-butynoate **V-5o**, the homocyclotrimerization product was observed [48].

In the reaction of **V-5m**, four new products of **V-1o**, **V-1′o**, **V-6o**, and **V-6′o** were isolated after hydrolysis in 2, 8, 4 and 24% yields, respectively (Scheme 5.4). The electronic-controlled regioselective alkyne insertion products **V-1o** and **V-6o** are the major products. The formation of reversed alkyne insertion species **V-1′o** and **V-6′o** may be due to the interaction between O and Ni atom in the reaction intermediates.

The regiochemical assignment of **V-1′o** was determined by the facts that there is no correlation between the cage C (74.2 ppm) and OCH$_2$ (3.84 ppm) can be observed in the HMBC analysis. On the other hand, the HMBC NMR spectrum of **V-1o** apparently illustrates the correlation between the proton on OCH$_2$ group (3.86 ppm) and the cage C (74.6 ppm) (Chart 5.1).

Chart 5.1 Assignment of the regioisomers of **V-1o** and **V-1′o**

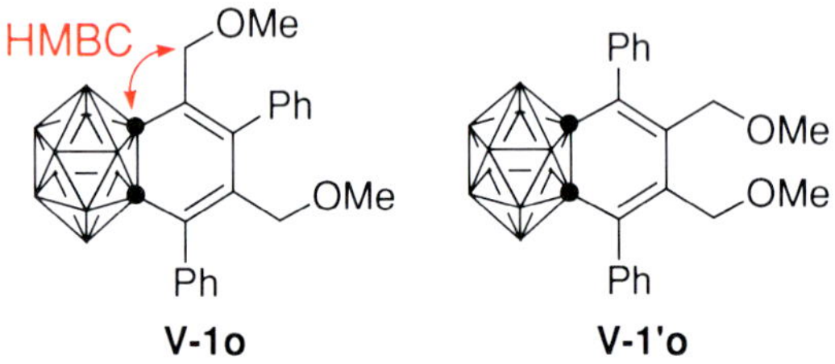

The relative regiochemical assignments of **V-6o** and **V-6′o** were determined using HH COSY analyses and the diagnostic correlation is shown in Chart 5.2.

Chart 5.2 Assignment of the regioisomers of **V-6o** and **V-6′o**

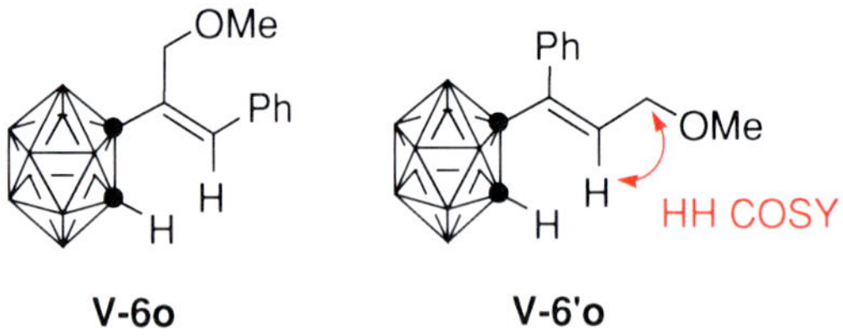

Internal diynes **V-7a-c** were also compatible with these nickel-catalyzed cyclo-addition reactions and gave the desired products **V-8** in 15–39% isolated yields with a good fused-ring size tolerance (Scheme 5.5). The yield was rather low for seven-

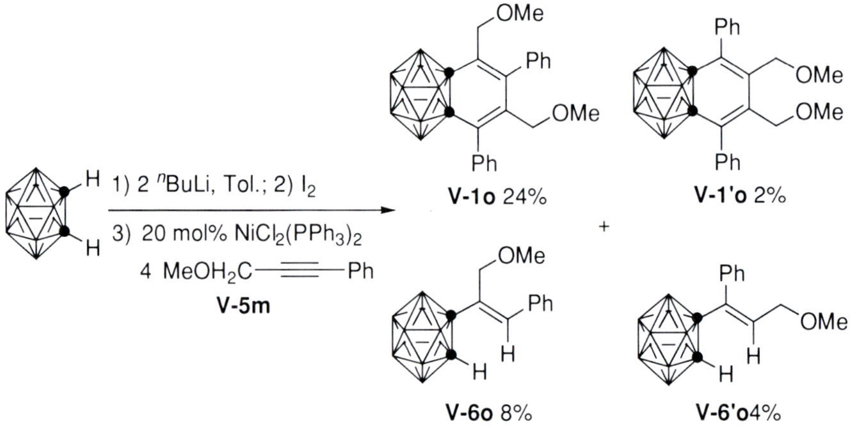

Scheme 5.4 The nickel-catalyzed cycloaddition of 1,2-*o*-carborynes with **V-5m**

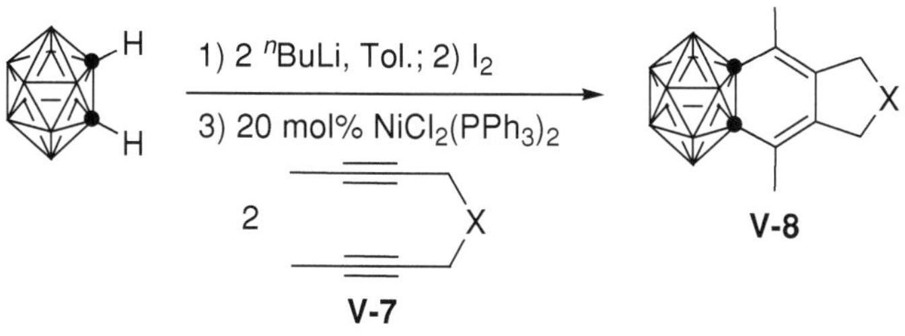

5.2 Results and Discussion 77

Scheme 5.5 Nickel-catalyzed cycloaddition of 1,2-o-carboryne with diynes

membered fused-ring species **V-8c**. No reaction proceeded for the oxo-bridged diyne **V-7d**. It's noteworthy that this condition is not suitable for the reaction involving alkenes and produces the coupling product in very low yield.

Compounds **V-1**, **V-6**, and **V-8** were fully characterized by ^{1}H, ^{13}C, and ^{11}B NMR spectra as well as high-resolution mass spectrometry. For products **V-1** without substituent on the B atom, the ^{11}B{^{1}H} NMR spectra generally exhibited a 2:6:2 or a 2:4:4 splitting pattern. And the ^{11}B{^{1}H} NMR spectra exhibited a 4:2:2:1:1 splitting pattern for **V-1b** and a 1:2:3:3:1 splitting pattern for **V-1c**. A singlet assignable to B-Ph in **V-1c** can be observed in the ^{1}H coupled ^{11}B NMR spectrum, whereas the signal of B–Cl is overlapped with other B–H signals and cannot be identified.

The molecular structures of **V-1h**, **V-1n**, **V-1o**, **V-6o** and **V-8b** were further confirmed by single-crystal X-ray analyses (Figs. 5.1, 5.2, 5.3, 5.4, 5.5). The localized double bonds suggest there is no π-delocalization in the six-membered ring of the benzocarborane products (Table 5.3).

To gain some insight into the reaction mechanism, an NMR reaction of 1-I-2-Li-1,2-$C_2B_{10}H_{10}$ with 1 equiv of Ni(cod)$_2$/2PPh$_3$ in toluene was conducted and monitored by ^{11}B and ^{31}P NMR spectra. The results suggested the formation of $(\eta^2$-$C_2B_{10}H_{10})$Ni(PPh$_3$)$_2$ even at room temperature, which indicates that an oxidative addition of I–C$_{cage}$ bond to Ni(0) proceeded. On the other hand, treatment of in situ generated 1-I-2-Li-1,2-$C_2B_{10}H_{10}$ with 1 equiv of NiCl$_2$(PPh$_3$)$_2$ in the presence of 2 equiv of n-butyl-2-pyridinylacetylene in refluxing toluene gave, after recrystallization from THF, a mono alkyne insertion product **V-9** [{[2-C(^{n}Bu)=C(o-C$_5$H$_4$N)-1,2-$C_2B_{10}H_{10}$]Ni}$_2$(μ_2-Cl)][Li(THF)$_4$] as red crystals in 25% yield (Scheme 5.6). It was fully characterized by various NMR spectra and elemental analyses.

Single-crystal X-ray analyses revealed that **V-9** is an ionic complex consisting of dimeric complex anions and tetrahedral cations. In the anion, two square-planar Ni moieties share one μ-Cl atom (Fig. 5.6). Coordination of the pyridinyl to the Ni

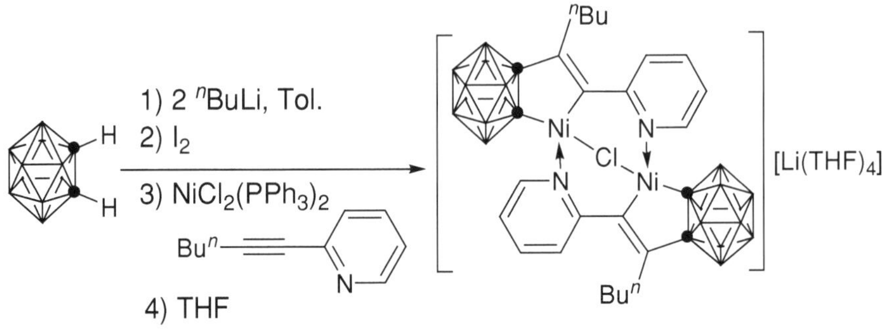

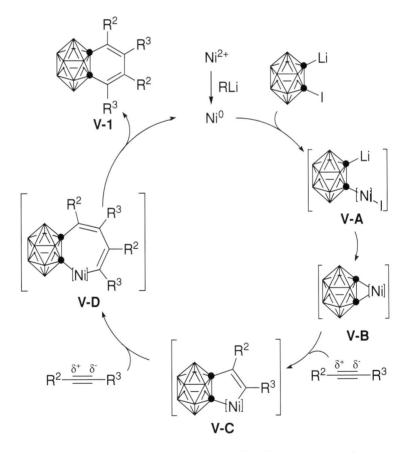

Scheme 5.6 Reaction of Ni-1,2-o-carboryne with n-butyl-2-pyridinylacetylene

Scheme 5.7 Proposed mechanism of nickel-catalyzed [2+2+2] cyclization reaction

5.2 Results and Discussion

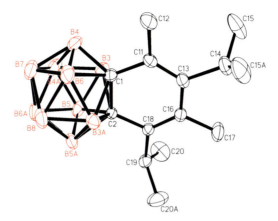

Fig. 5.1 Molecular structure of **V-1h**

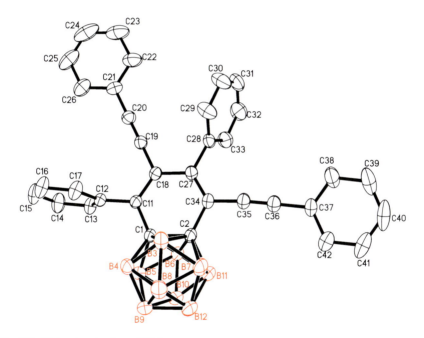

Fig. 5.2 Molecular structure of **V-1n**

atom can stabilize complex **V-9** and prevent the further insertion of the second equiv of *n*-butylpyridinylacetylene.

Given the above experimental evidence, a plausible mechanism for the nickel-catalyzed cycloaddition is shown in Scheme 5.7. The catalysis is likely initiated by Ni(0) species generated via the reduction of Ni(II) with lithiocarborane salt [49, 50]. Oxidative addition between I–C(cage) bond and Ni(0), followed by a subsequent elimination of lithium iodide produces a Ni-1,2-*o*-carboryne

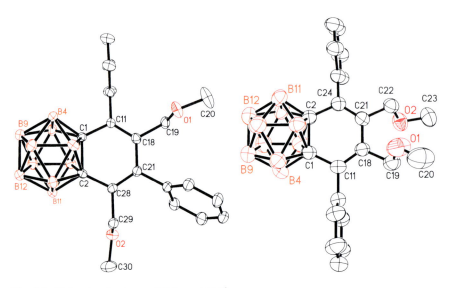

Fig. 5.3 Molecular structures of **V-1o** and **V-1'o**

Fig. 5.4 Molecular structure of **V-6o**

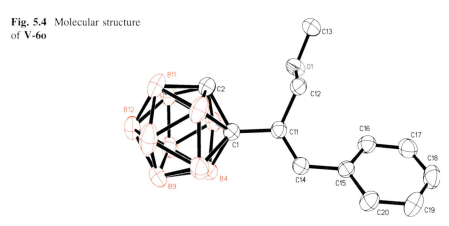

Fig. 5.5 Molecular structure of **V-8b**

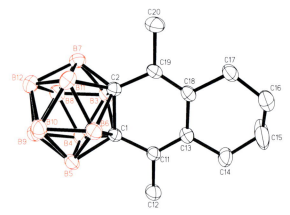

5.2 Results and Discussion

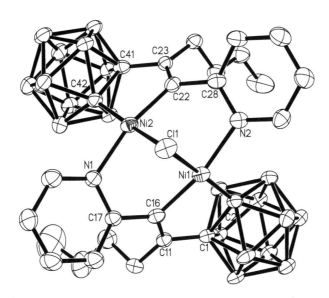

Fig. 5.6 Molecular structure of the anion in **V-9**. Selected bond lengths (Å) and angles (deg): Ni1–C2 1.890(7), Ni1–C16 1.929 (8), Ni1–Cl1 2.267(2), Ni1–N2 1.965(6), C1–C2 1.655(9), C1–C11 1.487(9), C11–C16 1.378(9), Ni2–C42 1.910(7), Ni2–C22 1.925(8), Ni2–Cl1 2.267(2), Ni2–N1 1.946(6), C41–C42 1.640 (10), C41–C23 1.507 (13), C23–C22 1.346 (10), C2–Ni1–C16 86.8(3), C16–Ni1–Cl1 95.7(2), Cl1–Ni1–N2 83.8(2), N2–Ni1–C2 96.8(3), C42–Ni2–C22 85.6(3), C22–Ni2–Cl1 97.1(2), Cl1–Ni2–N1 82.7(2), N1–Ni2–C42 97.6(3), Ni1–Cl1–Ni2 70.8(1)

Table 5.3 Selected bond distances (Å) for **V-1h**, **V-1n**, **V-1o**, **V-1′o** and **V-8b**

V-1h		V-1n		V-1o		V-1′o		V-8b	
C1–C2	1.629(5)	C1–C2	1.641(3)	C1–C2	1.639(6)	C1–C2	1.637(6)	C1–C2	1.643(4)
C1–C11	1.494(5)	C1–C11	1.486(3)	C1–C11	1.497(6)	C1–C11	1.502(6)	C1–C11	1.487(5)
C11–C13	1.347(5)	C11–C18	1.354(3)	C11–C18	1.350(6)	C11–C18	1.342(5)	C11–C13	1.346(5)
C13–C16	1.477(5)	C18–C27	1.473(3)	C18–C21	1.467(6)	C18–C21	1.460(6)	C13–C18	1.470(5)
C16–C18	1.346(5)	C27–C34	1.350(3)	C21–C28	1.347(6)	C21–C24	1.331(5)	C18–C19	1.343(5)
C18–C2	1.488(5)	C34–C2	1.496(3)	C28–C2	1.498(6)	C24–C2	1.508(6)	C19–C2	1.486(5)

intermediate **V-B**. An alternative pathway proceeded by the elimination of lithium iodide to form 1,2-*o*-carboryne and subsequent coordination to the metal center cannot be ruled out. Insertion of the first equiv of alkyne into the Ni–C(cage) bond of Ni-1,2-*o*-carboryne gives a nickelacyclopentene intermediate **V-C**. The second equiv of alkyne inserts into the Ni–C(vinyl) bond to afford the seven-membered intermediate **V-D** [45, 51–54]. Reductive elimination yields the cycloaddition product **V-1** and releases Ni(0) species to complete the catalytic cycle. The regioselectivity observed in the reactions can be rationalized by the polarity of alkynes [46, 47].

5.3 Summary

We have developed the first metal-catalyzed reaction of 1,2-*o*-carboryne with unsaturated molecules using 1-iodo-2-lithiocarborane as precursor and $NiCl_2(PPh_3)_2$ as catalyst. The mechanism was proposed after the structural confirmation of the key intermediate, nickelacyclopentene.

References

1. Gingrich HL, Ghosh T, Huang Q, Jones M Jr (1990) J Am Chem Soc 112:4082
2. Ghosh T, Gingrich HL, Kam CK, Mobraaten ECM, Jones M Jr (1991) J Am Chem Soc 113:1313
3. Huang Q, Gingrich HL, Jones M Jr (1991) Inorg Chem 30:3254
4. Cunningham RT, Bian N, Jones M Jr (1994) Inorg Chem 33:4811
5. Ho DM, Cunningham RJ, Brewer JA, Bian N, Jones M Jr (1995) Inorg Chem 34:5274
6. Barnett-Thamattoor L, Zheng G, Ho DM, Jones M Jr, Jackson JE (1996) Inorg Chem 35:7311
7. Atkins JH, Ho DM, Jones M Jr (1996) Tetrahedron Lett 37:7217
8. Lee T, Jeon J, Song KH, Jung I, Baik C, Park K-M, Lee SS, Kang SO, Ko J (2004) Dalton Trans 933
9. Jeon J, Kitamura T, Yoo B-W, Kang SO, Ko J (2001) Chem Commun 2110
10. Sayler AA, Beall H, Sieckhaus JF (1973) J Am Chem Soc 95:5790
11. Deng L, Chan H-S, Xie Z (2006) J Am Chem Soc 128:7728
12. Schore NE (1988) Chem Rev 88:1081
13. Trost BM (1991) Science 254:1471
14. Lautens M, Klute W, Tam W (1996) Chem Rev 96:49
15. Saito S, Yamamoto Y (2000) Chem Rev 100:2901
16. Buchwald SL, Nielsen RB (1988) Chem Rev 88:1047
17. Bennett MA, Schwemlein HP (1989) Angew Chem Int Ed Engl 28:1296
18. Bennett MA, Wenger E (1997) Chem Ber 130:1029
19. Jones WM, Klosin J (1998) Adv Organomet Chem 42:147
20. Retbøll M, Edwards AJ, Rae AD, Willis AC, Bennett MA, Wenger E (2002) J Am Chem Soc 124:8348
21. Peña D, Escudero S, Pérez D, Guitián E, Castedo L (1998) Angew Chem Int Ed 37:2659
22. Peña D, Pérez D, Guitián E, Castedo L (1999) Org Lett 1:1555
23. Romero C, Peña D, Pérez D, Guitián E (2006) Chem Eur J 12:5677
24. Peña D, Pérez D, Guitián E, Castedo L (1999) J Am Chem Soc 121:5827
25. Radhakrishnan KV, Yoshikawa E, Yamamoto Y (1999) Tetrahedron Lett 40:7533
26. Peña D, Pérez D, Guitián E, Castedo L (2000) J Org Chem 65:6944
27. Sato Y, Tamura T, Mori M (2004) Angew Chem Int Ed 43:2436
28. Romero C, Peña D, Pérez D, Guitián E (2008) J Org Chem 73:7996
29. Yoshikawa E, Radhakrishnan KV, Yamamoto Y (2000) J Am Chem Soc 122:7280
30. Yoshikawa E, Yamamoto Y (2000) Angew Chem Int Ed 39:173
31. Quintana I, Boersma AJ, Peña D, Pérez D, Guitián E (2006) Org Lett 8:3347
32. Jeganmohan M, Cheng C-H (2004) Org Lett 6:2821
33. Jayanth TT, Jeganmohan M, Cheng C-H (2005) Org Lett 7:2921
34. Henderson JL, Edwards AS, Greaney MF (2006) J Am Chem Soc 128:7426
35. Bhuvaneswari S, Jeganmohan M, Yang M-C, Cheng C-H (2008) Chem Commun 2158
36. Xie C, Liu L, Zhang Y, Xu P (2008) Org Lett 10:2393
37. Bhuvaneswari S, Jeganmohan M, Cheng C-H (2008) Chem Commun 5013

References

38. Jeganmohan M, Bhuvaneswari S, Cheng C-H (2009) Angew Chem Int Ed 48:391
39. Henderson JL, Edwards AS, Greaney MF (2007) Org Lett 9:5589
40. Liu Z, Zhang X, Larock RC (2005) J Am Chem Soc 127:15716
41. Liu Z, Larock RC (2007) Angew Chem Int Ed 46:2535
42. Bhuvaneswari S, Jeganmohan M, Cheng C-H (2006) Org Lett 8:5581
43. Hsieh J-C, Cheng C-H (2005) Chem Commun 2459
44. Jayanth TT, Cheng C-H (2007) Angew Chem Int Ed 46:5921
45. Bennett MA, Glewis M, Hockless DCR, Wenger E (1997) J Chem Soc Dalton Trans 3105
46. Bennett MA, Macgregor SA, Wenger E (2001) Helv Chim Acta 84:3084
47. Deng L, Chan H-S, Xie Z (2006) J Am Chem Soc 128:7728
48. Jhingan AK, Maier WF (1987) J Org Chem 52:1161
49. Miyaura M (2002) Cross-coupling reactions. A practical guide in topics in current chemistry. Springer, Berlin
50. Terao J, Tomita M, Singh SP, Kambe N (2010) Angew Chem Int Ed 49:144
51. Xie Z (2006) Coord Chem Rev 250:259
52. Sun Y, Chan H-S, Zhao H, Lin Z, Xie Z (2006) Angew Chem Int Ed 45:5533
53. Sun Y, Chan H-S, Xie Z (2006) Organometallics 25:3447
54. Shen H, Xie Z (2009) Chem Commun 2431

Chapter 6
Palladium/Nickel-Cocatalyzed [2+2+2] Cycloaddition of 1,3-*o*-Carboryne with Alkynes

6.1 Introduction

1,2-*o*-Carboryne is a three-dimentional relative of benzyne (Chart 6.1) [1]. It can react with alkenes, dienes, and alkynes in [2+2], [2+4] cycloaddition and ene-reaction patterns [2–10] similar to those of benzyne [11–16]. This reactive species can be stabilized by transition metals, leading to the formation of metal-1,2-*o*-carboryne complexes [17–22]. Molecular orbital calculations on the Zr-carboryne complex suggest that the bonding interactions between Zr and carboryne are best described as a resonance hybrid of both Zr–C σ and Zr–C π bonding forms [21], which is similar to that observed in Zr-benzyne complex (Chart 6.1) [23]. This type of metal-1,2-*o*-carboryne complexes can react with unsaturated molecules in a control manner to produce alkenylcarboranes, benzocarboranes [24], dihydrobenzocarboranes, and other functionalized carboranes [25–27]. In view of these unique features of 1,2-*o*-carboryne and the important application of boron-centered nucleophiles [28–41], we became interested in the unknown species 1,3-dehydro-*o*-carborane (1,3-*o*-carboryne) (Chart 6.1). In this section, we report our work on palladium/nickel-cocatalyzed reaction of 1,3-*o*-carboryne with 2 equiv of alkynes to afford [2+2+2] cycloaddition products, 1,3-benzo-*o*-carborane.

Chart 6.1 Structures of benzyne and carborynes

6.2 Results and Discussion

6.2.1 Synthesis of 1,3-o-Carboryne Precursor

As 1,2-o-carboryne can be generated in situ by heating 1-X-2-Li-1,2-$C_2B_{10}H_{10}$ (X = Br [2–8], I) via the elimination of LiX, we attempted to produce 1,3-o-carboryne in a similar manner using 1-Li-3-X-1,2-$C_2B_{10}H_{10}$ as precursors. Unfortunately, both 1-Li-3-X-1,2-$C_2B_{10}H_{10}$ and 1-Li-2-CH_3-3-X-1,2-$C_2B_{10}H_9$ are all very thermally stable even after prolonged heating in THF or toluene. Considering that the cage B–I bond can undergo oxidative addition in the presence of Pd(0) [42–45], we speculate that an oxidative addition of the cage B–I in 1-Li-3-I-1,2-$C_2B_{10}H_{10}$ on the Pd(0), followed by subsequent elimination of LiI would afford the target complex Pd-1,3-o-carboryne, which could be trapped by alkynes.

A series 1,3-o-carboryne precursors **VI-1a–c** were synthesized by the boron insertion reaction of the dicarbollide anion $C_2B_9H_{11}^{2-}$ with BI_3 [46–51]. Compounds **VI-1d–g** can be synthesized with lithiated **VI-1a** and alkyl chloride (Scheme 6.1).

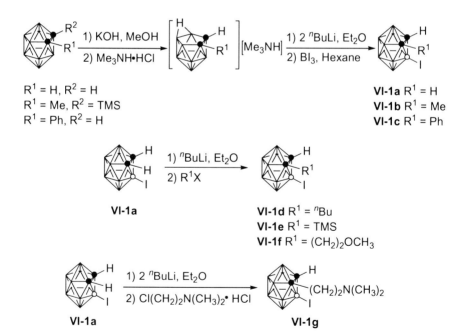

Scheme 6.1 Synthesis of 3-iodo-1,2-o-carboranes

6.2 Results and Discussion

Table 6.1 Reaction of 1,3-*o*-carboryne precursor **VI-1b**[a]

Entry	Catalyst	Loading (mol%)	Reaction time (h)	Yield (%)[b]	
				VI-1b	VI-3b
1	Pd(PPh$_3$)$_4$	10	14	<1	>99
2	Pd(PPh$_3$)$_4$	5	30	10	90
3	Pd(PPh$_3$)$_4$/Ni(cod)$_2$	5/5	30	<1	>99
4	Ni(cod)$_2$	5	30	>99	<1

[a] Conditions: (1) ^{n}BuLi (1 equiv), toluene, r.t., 0.5 h; (2) Catalyst, 110 °C
[b] Yields determined by GC–MS on the crude product mixture

6.2.2 Reaction of 1,3-o-Carboryne Precursor

The reactions of these precursors were next studied. In an initial attempt, a toluene solution of the 1-Li-2-Me-3-I-1,2-C$_2$B$_{10}$H$_9$, prepared in situ by treatment of 2-Me-3-I-1,2-C$_2$B$_{10}$H$_{10}$ (**VI-1b**) with 1 equiv of ^{n}BuLi, was heated in the presence of Pd(PPh$_3$)$_4$ (10 mol%) to give 1-methyl-*o*-carborane (**VI-3b**) in almost quantitative yield in 14 h. The formation of **VI-3b** may probably result from the decomposition of Pd-2-methyl-1,3-*o*-carboryne at high temperatures (Table 6.1, entry 1). If the catalyst loading was reduced to 5 mol%, the reaction was slow down (Table 6.1, entry 2). Ni(cod)$_2$ was almost inactive in the activation of cage B–I bond (Table 6.1, entry 4). However a combination of 5 mol% of Ni(cod)$_2$ and 5 mol% Pd(PPh$_3$)$_4$ can improve the formation of **VI-3b** (Table 6.1, entry 3) (vide infra) [52]. Grinard reagent (MeMgBr) is less effective than ^{n}BuLi in the reaction with cage CH. On the other hand, MeMgBr can react with 3-iodo-*o*-carborane in the presence of Pd(0) to give 3-methyl-*o*-carborane [53–57]. 3-Bromo-*o*-carborane and 3-chloro-*o*-carborane are not suitable for this reaction because Pd(0) cannot add to the boron-bromine or boron-chlorine bond efficiently [56]. It is noted that no reaction proceeded at $T < 70$ °C, and only compound **VI-3b** can be observed at higher temperatures by ^{11}B NMR spectroscopy in the reaction of 3-iodo-1-lithio-2-methyl-*o*-carborane with a catalytic amount of Pd(PPh$_3$)$_4$. Attempts to isolate (η^2-1,3-*o*-C$_2$B$_{10}$H$_{10}$)Pd(L), an analogue of (η^2-1,2-*o*-C$_2$B$_{10}$H$_{10}$)Ni(L) [17], in the presence of PPh$_3$ or dppe (dppe = 1,2-bis(diphenylphosphino)ethane) failed.

88   6 Palladium/Nickel-Cocatalyzed [2+2+2] Cycloaddition

Table 6.2 Optimization of Pd/Ni-catalyzed cycloaddition reaction[a]

Entry	Catalyst	Loading (mol%)	Reaction time	Yield[b] VI-1b	VI-3b	VI-4b
1	None	0	7 d[c]	100	–	–
2	Pd(OAc)$_2$	10	3 d[c]	56	19	25
3	PdCl$_2$(PPh$_3$)$_2$	10	3 d[c]	9	12	79
4	PdCl$_2$(cod)	10	3 d[c]	87	12	<1
5	PdCl$_2$(cod)/2PPh$_3$	10	30 h	<1	8	91
6	Pd(CH$_2$TMS)$_2$(cod)	10	3 d[c]	90	9	<1
7	[Pd(Ally)Cl]$_2$	5	3 d[c]	6	25	69
8	[Pd(Ally)Cl]$_2$/4PPh$_3$	5	1 h	<1	7	92
9	Pd(dba)$_2$	10	7 d[c]	21	36	43
10	Pd(PPh$_3$)$_4$	10	7 h	2	8	90
11	Ni(cod)$_2$	10	7 d[c]	72	13	15
12	Pd(PPh$_3$)$_4$/Ni(PPh$_3$)$_4$	10/10	3 h	3	6	91
13	Pd(dba)$_2$/Ni(cod)$_2$	10/10	3 d[c]	25	43	32
14	Pd(PPh$_3$)$_4$/Ni(cod)$_2$	10/10	0.5 h	<1	6	93
15	Pd(PPh$_3$)$_4$/Ni(cod)$_2$	5/5	2 h	<1	3	96
16	Pd(PPh$_3$)$_4$/Ni(cod)$_2$/2PPh$_3$	5/5	2 h	<1	4	95
17	Pd(PPh$_3$)$_4$/Ni(cod)$_2$	2/2	4 h	<1	6	93
18	PdCl$_2$(PPh$_3$)$_2$/Ni(cod)$_2$	10/10	3 h	9	7	84

[a] Conditions: (1) ^{n}BuLi (1 equiv), toluene, r.t., 0.5 h; (2) Catalyst, 3-hexyne (4 equiv), 110 °C
[b] Yields determined by GC–MS on the crude product mixture
[c] The reaction was quenched with H$_2$O

6.2.3 Metal-Catalyzed [2+2+2] Cycloaddition of 1,3-o-Carboryne with Alkynes

Subsequent work focused on trapping the 1,3-o-carboryne intermediate with alkynes. The optimization of this reaction is listed in Table 6.2. Pd(II) species can effectively catalyze the [2+2+2] cycloaddition reaction of 2-methyl-1,3-o-carbo-ryne with 3-hexyne to afford **VI-4b** (entries 2–8). Adding PPh$_3$ to PdCl$_2$(cod) or [Pd(Ally)Cl]$_2$ led to a big increase in the isolation of **VI-4b** probably due to the reduction of Pd(II) to Pd(0) by PPh$_3$ (entries 5 and 8) [58]. Pd(0) species is more effective but with a big ligand effect. Pd(PPh$_3$)$_4$ can catalyze the [2+2+2] cycloaddition reaction affording **VI-4b** in 90% yield. In comparison, Pd(dba)$_2$ (dba = dibenzylideneacetone) gives **VI-4b** in 43% yield only (Table 6.2, entries 9

6.2 Results and Discussion

Table 6.3 Pd/Ni-Catalyzed cycloaddition of 1,3-o-carboryne with alkynes[a]

Entry	R^1/**VI-1**	R^2/R^3/**VI-2**	Product	Yield (%)[b]
1	H/**VI-1a**	Et/Et/**VI-2a**	**VI-4a**	12
2	Me/**VI-1b**	Et/Et/**VI-2a**	**VI-4b**	79
3	^{n}Bu/**VI-1c**	Et/Et/**VI-2a**	**VI-4c**	67
4[c]	TMS/**VI-1d**	Et/Et/**VI-2a**	**VI-4d**	69
5	Ph/**VI-1e**	Et/Et/**VI-2a**	**VI-4e**	43
6	(CH$_2$)$_2$OMe/**VI-1f**	Et/Et/**VI-2a**	**VI-4f**	58
7	(CH$_2$)$_2$NMe$_2$/**VI-1g**	Et/Et/**VI-2a**	**VI-4g**	51
8	Me/**VI-1b**	^{n}Pr/^{n}Pr/**VI-2b**	**VI-4h**	55
9	Me/**VI-1b**	^{n}Bu/^{n}Bu/**VI-2c**	**VI-4i**	43
10	Me/**VI-1b**	^{n}Bu/TMS/**VI-2d**	NR[d]	–
11	Me/**VI-1b**	COOMe/COOMe/**VI-2e**	NR[e]	–
12	Me/**VI-1b**	Me/COOMe/**VI-2f**	NR[e]	–
13	Me/**VI-1b**	Ph/Ph/**VI-2g**	**VI-4j**	55
14	Me/**VI-1b**	p-Tolly/p-Tolly/**VI-2h**	**VI-4k**	51
15	Me/**VI-1b**	o-Tolly/o-Tolly/**VI-2i**	NR[d]	–
16	Me/**VI-1b**	Me/Ph/**2j**	**VI-4l** **VI-5l**	49 (**VI-4l/VI-5l** = 62/38)[f]
17	Me/**VI-1b**	Et/Ph/**VI-2k**	**VI-4m** **VI-5m**	47(**VI-4m/VI-5m** = 80/20)[f]

[a] Conditions: (1) ^{n}BuLi (1 equiv), toluene, r.t., 0.5 h; (2) Pd(PPh$_3$)$_4$ (5 mol%), Ni(cod)$_2$ (5 mol%), alkyne (4 equiv), 110 °C, overnight
[b] Isolated yields
[c] 5 mol% Pd(PPh$_3$)$_4$ used as catalyst
[d] **VI-3** was obtained as the product
[e] **VI-1** was recovered
[f] Ratio was determined by ^{1}H NMR spectroscopy on the crude product mixture

and 10). Ni(cod)$_2$ exhibited very low catalytic activity (Table 6.2, entry 11), but the addition of nickel species to palladium catalyst can significantly accelerate the reaction (entries 12–18) [52]. Combination of Pd(PPh$_3$)$_4$ with Ni(cod)$_2$ exhibited the highest catalytic activity in this [2+2+2] cyclization. The similar results were observed when the catalyst loading was decreased from 10 to 2 mol% or 2 equiv of PPh$_3$ was add in the reaction (entries 14–17). Ni(cod)$_2$ can also accelerate the catalytic reaction of PdCl$_2$(PPh$_3$)$_2$ giving **4b** in 84% yield in 3 h (entry 18).

Listed in Table 6.3 are representative results obtained from the palladium/nickel-cocatalyzed cycloaddition reactions with various alkynes. 1,3-o-Carboryne without protecting group at 2-C position gives very low isolated yield (12%) of **VI-4a** (entry 1). The steric factor on the 2-C of 1,3-*o*-carboryne has no significant effect on the reactions (entries 2–4). Functionalized 1,3-*o*-carboryne with electron-withdrawing group such as phenyl leading to the isolation of corresponding cyclization product **VI-4e** in moderate yield (entry 5). And substituents bearing heteroatom also afford moderate yields (entries 6 and 7). Both aliphatic and aromatic alkynes underwent [2+2+2] cycloaddition reactions. Steric factors played an important role in the reactions. No reaction proceeded for sterically more demanding alkynes bearing trimethylsilyl or *o*-tolyl group (entries 10 and 15). Alkynes bearing carbonyl group such as **VI-2e** and **VI-2f** were incompatible with this reaction because they can react with the carboryne precursor 1-Li-2-Me-3-I-1,2-$C_2B_{10}H_9$ (entries 11 and 12). Unsymmetrical alkynes gave two isomers of **VI-4l,m** and **VI-5l,m** (entries 16 and 17). It is noted that alkenes only give trace amount insertion products, and nitriles, isonitriles, or carbodiimides are ineffective under this condition.

The palladium/nickel-cocatalyzed cycloaddition reaction was successfully extended to various diynes (Scheme 6.2). Thus, 2-methyl-1,3-*o*-carboryne underwent cycloaddition with diynes (**VI-6a–d**) to provide the 1,3-benzo-*o*-carborane products, **VI-7a–c** in 6–34% yields, with a good fused-ring size tolerance.

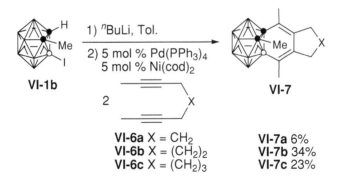

Scheme 6.2 Pd/Ni-Cocatalyzed cycloaddition of 1,3-*o*-carboryne with diynes

Compounds **VI-4**, **VI-5** and **VI-7** were fully characterized by ^{1}H, ^{13}C, and ^{11}B NMR spectra as well as high-resolution mass spectrometry. For compounds **VI-4a–g**, four triplets at 0.5–1.2 ppm and a multiplet at ∼2.5 ppm corresponding to the ethyl groups can observed in the ^{1}H NMR spectra. The C*H*₃ (cage) signals appear at 1.2 ppm in the ^{1}H NMR spectra of **VI-4b,h,i** and **VI-7a–c**. In case of phenyl substituted products **VI-4j,k** the C*H*₃(cage) signals were shifted lowfield to 2.1 ppm. In the ^{1}H NMR spectra of **VI-4l,m** and **VI-5l,m**, which have both alkyl and aryl substitutents, the C*H*₃(cage) signals were observed at 1.7 ppm. Their ^{13}C NMR spectra were consistent with the ^{1}H NMR results. The olefin carbons which connected to the boron atom were not observed for **VI-4**, **VI-5** and **VI-7** [59]. The ^{11}B{^{1}H} NMR spectra generally exhibited a 3:5:2 splitting pattern for **VI-4b,h–m**,

6.2 Results and Discussion

Fig. 6.1 Molecular structure of **VI-4a**

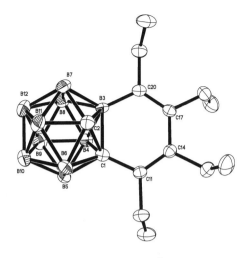

Fig. 6.2 Molecular structure of **VI-4b**

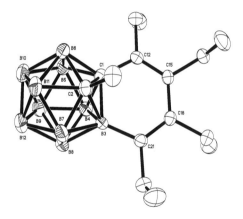

VI-5m and **VI-7a–c**, bearing methyl groups on the cage carbon. And the $^{11}B\{^1H\}$ NMR spectra displayed a 1:2:1:3:1:1:1, 3:4:1:1:1, 1:4:3:1:1, 2:1:1:3:3, 1:1:5:1:1:1, and 3:5:1:1 pattern for **VI-4a**, **4c**, **4d**, **4e**, **4f** and **4g**, respectively. The signal of *B*–*C* is overlapped with other *B*–H signals and cannot be identified.

The molecular structures of **VI-4a**, **VI-4b**, **VI-4d**, **VI-4j**, **VI-4m**, **VI-5m** and **VI-7b** were further confirmed by single-crystal X-ray analyses (Figs. 6.1, 6.2, 6.3, 6.4, 6.5, 6.6). The localized double bonds suggest there is no substantial π-delocalization in the six-membered ring (Table 6.4).

Fig. 6.3 Molecular structure of **VI-4d**

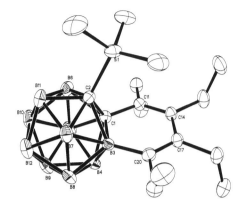

Fig. 6.4 Molecular structure of **VI-4j**

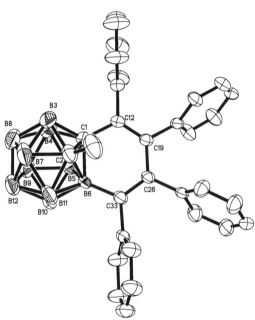

6.2.4 Proposed Mechanism

It is believed that the reaction is through a metal-1,3-*o*-carboryne intermediate because the catalytic amount of Pd species can convert **VI-1b** to **VI-3b** quantitatively. A mixture of **VI-1b** and 3-hexyne was refluxed in toluene in the presence of 5 mol% Pd(PPh$_3$)$_4$ and 5 mol% Ni(cod)$_2$ did not give any alkyne insertion products, rather afforded the isomers of **VI-1b** with iodo being located at different cage boron positions as suggested by GC–MS analyses.

6.2 Results and Discussion

Fig. 6.5 Molecular structures of **VI-4m** and **VI-5m**

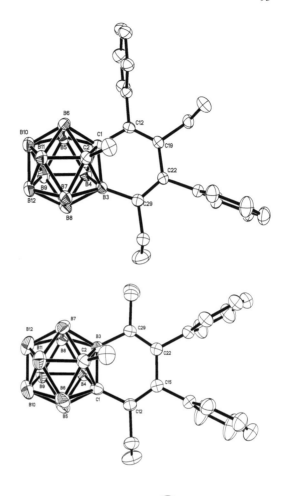

Fig. 6.6 Molecular structure of **VI-7b**

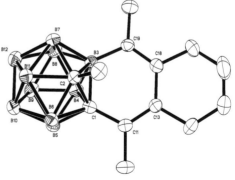

Table 6.4 Selected bond lengths (Å)

VI-4a	C(1)–B(3)	C(1)–C(11)	C(11)–C(14)	C(14)–C(17)	C(17)–C(20)	C(20)–B(3)
	1.706(5)	1.507(4)	1.340(4)	1.482(4)	1.351(4)	1.524(5)
VI-4b	C(1)–B(3)	C(1)–C(12)	C(12)–C(15)	C(15)–C(18)	C(18)–C(21)	C(21)–B(3)
	1.708(4)	1.502(4)	1.353(4)	1.479(4)	1.351(4)	1.527(5)
VI-4d	C(1)–B(3)	C(1)–C(11)	C(11)–C(14)	C(14)–C(17)	C(17)–C(20)	C(20)–B(3)
	1.707(2)	1.518(2)	1.366(3)	1.498(2)	1.366(3)	1.546(3)
VI-4j	C(1)–B(6)	C(1)–C(12)	C(12)–C(19)	C(19)–C(26)	C(26)–C(33)	C(33)–B(6)
	1.706(4)	1.520(4)	1.347(3)	1.487(3)	1.357(3)	1.527(4)
VI-4m	C(1)–B(3)	C(1)–C(12)	C(12)–C(19)	C(19)–C(22)	C(22)–C(29)	C(29)–B(3)
	1.707(5)	1.481(5)	1.350(4)	1.481(5)	1.344(5)	1.537(6)
VI-5m	C(1)–B(3)	C(1)–C(12)	C(12)–C(15)	C(15)–C(22)	C(22)–C(29)	C(29)–B(3)
	1.701(4)	1.514(4)	1.352(4)	1.492(4)	1.351(4)	1.531(4)
VI-7b	C(1)–B(3)	C(1)–C(11)	C(11)–C(13)	C(13)–C(18)	C(18)–C(19)	C(19)–B(3)
	1.689(5)	1.508(5)	1.359(5)	1.483(5)	1.356(5)	1.533(5)

A plausible mechanism for palladium/nickel-cocatalyzed [2+2+2] cocyclization is shown in Scheme 6.3. As Ni(0) cannot insert into the B–I bond efficiently (Table 6.1, entry 4), the Pd-1,3-*o*-carboryne **VI-B** is formed by the oxidative addition of B–I on Pd(0), followed by LiI elimination. It is noteworthy that the reactions were very slow (>5 days) and inefficient with more bulky alkynes **VI-2b–k** when only Pd(PPh$_3$)$_4$ or [Pd(Ally)Cl]$_2$/PPh$_3$ was employed as catalyst. In view of that the two-component catalyst is more effective than Pd species alone in the reaction of 1,3-*o*-carboryne with alkynes, it is rational to propose a transmetallation process between Pd and Ni, affording a more reactive

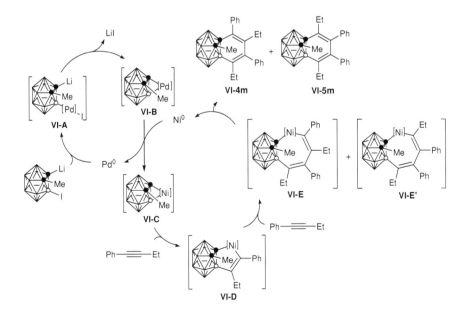

Scheme 6.3 Proposed mechanism of Pd/Ni-cocatalyzed [2+2+2] cyclization reaction

nickel-1,3-o-carboryne **VI-C**. The relatively higher activity of Ni species is probably due to that the Pd–B bond is stronger than the Ni–B bond or the Ni–B bonding pair is more nucleophilic than that of Pd–B. In the reaction with PhC≡CEt, the electronically controlled regio-selective insertion of unsymmetrical alkyne into the Ni–B bond gives the nickelacyclopentene intermediate **VI-D** [60, 61]. The absence of 2-Me-1,3-{1′,4′-[EtC=C(C_6H_5)-C(Et)=C(C_6H_5)]}-1,2-$C_2B_{10}H_{10}$ in the products indicates the exclusive insertion of Ni–B bond. As the insertion of alkynes into the Ni–C(cage) bond in metal-carboranyl complexes is prohibited due to steric reasons [62–65], the second equivalent of alkyne inserts into the Ni–C(vinyl) bond in both head-to-tail and head-to-head manners. Subsequent reductive elimination yields the final products **VI-4m** and **VI-5m**.

The M–B bond is much more reactive than the M–C bond in the alkyne insertion as the bonding pair of M–B is very high in energy. This is consistent with the result from the reaction of metal-borataalkene with alkynes [59] and the conclusion based on metal-catalyzed borylation reactions [66, 67]. Due to the low electronegativity of boron, an M–B bond is much more nucleophilic than an M–C bond. The alkyne insertion into an M–B bond step can be considered as a nucleophilic attack of the M–B σ-bond (the bonding electron pair) on one of the two alkyne carbons. The nucleophilic attack in nature also explains the regioselectivity observed in the unsymmetrical alkynes, i.e., in the insertion product **VI-D**, boron is bonded to the carbon having the electron-donating ethyl substituent.

6.3 Summary

In summary, we have shown for the first time a 1,3-o-carboryne, which can be regarded as a new boron nucleophile and can be trapped by unsaturated molecules in the presence of transition metal. This serves a palladium/nickel-cocatalyzed [2+2+2] cycloaddition reaction of 1,3-o-carboryne with alkynes to afford 1,3-benzo-o-carboranes. This work offers a new methodology for B-functionalization of carborane and demonstrates the relative reactivity of M–C over M–B bond in 1,3-o-carboryne complexes toward alkynes.

References

1. Kiran B, Anoop A, Jemmis ED (2002) J Am Chem Soc 124:4402
2. Gingrich HL, Ghosh T, Huang Q, Jones M Jr (1990) J Am Chem Soc 112:4082
3. Ghosh T, Gingrich HL, Kam CK, Mobraaten ECM, Jones M Jr (1991) J Am Chem Soc 113:1313
4. Huang Q, Gingrich HL, Jones M Jr (1991) Inorg Chem 30:3254
5. Cunningham RT, Bian N, Jones M Jr (1994) Inorg Chem 33:4811
6. Ho DM, Cunningham RJ, Brewer JA, Bian N, Jones M Jr (1995) Inorg Chem 34:5274

7. Barnett-Thamattoor L, Zheng G, Ho DM, Jones M Jr, Jackson JE (1996) Inorg Chem 35:7311
8. Atkins JH, Ho DM, Jones M Jr (1996) Tetrahedron Lett 37:7217
9. Lee T, Jeon J, Song KH, Jung I, Baik C, Park K-M, Lee SS, Kang SO, Ko J (2004) Dalton Trans 933–937
10. Jeon J, Kitamura T, Yoo B-W, Kang SO, Ko J (2001) Chem Commun 2110–2111
11. Roberts JD, Simmons HE Jr, Carlsmith LA, Vaughan CW (1953) J Am Chem Soc 75:3290
12. Hart H (1994) In: Patai S (ed) Chemistry of triple-bonded functional groups, supplement C2, Chap. 18. Wiley, Chichester
13. Hoffmann RW (1967) Dehydrobenzene and cycloalkynes. Academic Press, New York
14. Gilchrist TL (1983) In: Patai S, Rappoport Z (eds) Chemistry of functional groups, supplement C, Chap. 11. Wiley, Chichester
15. Buchwald SL, Nielsen RB (1988) Chem Rev 88:1047
16. Jones WM, Klosin J (1998) Adv Organomet Chem 42:147
17. Sayler AA, Beall H, Sieckhaus JF (1973) J Am Chem Soc 95:5790
18. Zakharkin LI, Kovredov AI (1975) Izv Akad Nauk SSSR, Ser Khim 2619
19. Ol'dekop YA, Maier NA, Erdman AA, Prokopovich VP (1981) Dokl Akad Nauk SSSR 257:647
20. Ol'dekop YA, Maier NA, Erdman AA, Prokopovich VP (1982) Zh Obshch Khim 52:2256
21. Wang H, Li H-W, Huang X, Lin Z, Xie Z (2003) Angew Chem Int Ed 42:4347
22. Ren S, Deng L, Chan H-S, Xie Z (2009) Organometallics 28:5749
23. Buchwald SL, Watson BT (1986) J Am Chem Soc 108:7411
24. Deng L, Chan H-S, Xie Z (2006) J Am Chem Soc 128:7728
25. Deng L, Chan H-S, Xie Z (2005) J Am Chem Soc 127:13774
26. Ren S, Chan H-S, Xie Z (2009) Organometallics 28:4106
27. Ren S, Chan H-S, Xie Z (2009) J Am Chem Soc 131:3862
28. Braunschweig H, Dewhurst RD, Schneider A. (2010) Chem Rev 110:3924
29. Mkhalid IAI, Barnard JH, Marder TB, Murphy JM, Hartwig JF (2010) Chem Rev 110:890
30. Hartwig JF (2008) Nature 455:314
31. Coombs DL, Aldridge S (2004) Coord Chem Rev 248:535
32. Braunschweig H, Colling M (2001) Coord Chem Rev 223:1
33. Irvine GJ, Gerald Lesley MJ, Marder TB, Norman NC, Rice CR, Robins EG, Roper WR, Whittell GR, Wright L (1998) J Chem Rev 98:2685
34. Braunschweig H (1998) Angew Chem Int Ed 37:1786
35. Wadepohl H (1997) Angew Chem Int Ed 36:2441
36. Segawa Y, Yamashita M, Nozaki K (2006) Science 314:113
37. Yamashita M, Suzuki Y, Segawa Y, Nozaki K (2007) J Am Chem Soc 129:9570
38. Segawa Y, Suzuki Y, Yamashita M, Nozaki K (2008) J Am Chem Soc 130:16069
39. Segawa Y, Yamashita M, Nozaki K (2009) J Am Chem Soc 131:9201
40. Terabayashi T, Kajiwara T, Yamashita M, Nozaki K (2009) J Am Chem Soc 131:14162
41. Spokoyny AM, Reuter MG, Stern CL, Ratner MA, Seideman T, Mirlin CA (2009) J Am Chem Soc 131:9482–9483
42. Li J, Logan CF, Jones M Jr (1991) Inorg Chem 30:4866
43. Zheng Z, Jiang W, Zinn AA, Knobler CB, Hawthorne MF (1995) Inorg Chem 34:2095
44. Jiang W, Knobler CB, Curtis CE, Mortimer MD, Hawthorne MF (1995) Inorg Chem 34:3491
45. Marshall WJ, Young RJ Jr, Grushin VV (2001) Organometallics 20:523
46. Hawthorne MF, Wegner PA (1965) J Am Chem Soc 87:4302
47. Hawthorne MF, Wegner PA (1968) J Am Chem Soc 90:896
48. Potapova TV, Mikhailov BM (1967) Izu Akad Nauk SSSR, Ser Khim, 2367
49. Roscoe JS, Kongpricha S, Papetti S (1970) Inorg Chem 9:1561
50. Li J, Jones M Jr (1990) Inorg Chem 29:4162
51. Li J, Caparrelli DJ, Jones M Jr (1993) J Am Chem Soc 115:408
52. Cho CS, Lee JW, Lee DY, Shim SC, Kim TJ (1996) Chem Commun 2115
53. Zakharkin LI, Kovredov AI, Ol'shevskaya VA, Shaugumbekova Zh S (1980) Izv Akad Nauk SSSR, Ser Khim 1691

References

54. Kovredov AI, Shaugumbekova ZS, Petrovskii PV, Zakharkin LI (1989) Zh Obshch Khrm 59:607
55. Jiang W, Knobler CB, Curtis CE, Mortimer MD, Hawthorne MF (1995) Inorg Chem 34:3491
56. Clara Vinas C, Barberà G, Oliva JM, Teixidor F, Welch AJ, Rosair GM (2001) Inorg Chem 40:6555
57. Mukhin SN, Kabytaev KZ, Zhigareva GG, Glukhov IV, Starikova ZA, Bregadze VI, Beletskaya IP (2008) Organometallics 27:5937
58. Miyaura M (2002) Cross-coupling reactions. A practical guide in topics in current chemistry. Springer, Berlin
59. Cook KS, Piers WE, McDonald R (2002) J Am Chem Soc 124:5411
60. Bennett MA, Glewis M, Hockless DCR, Wenger E (1997) J Chem Soc Dalton Trans 3105
61. Bennett MA, Macgregor SA, Wenger E (2001) Helv Chim Acta 84:3084
62. Xie Z (2006) Coord Chem Rev 250:259
63. Sun Y, Chan H-S, Zhao H, Lin Z, Xie Z (2006) Angew Chem Int Ed 45:5533
64. Sun Y, Chan H-S, Xie Z (2006) Organometallics 25:3447
65. Shen H, Xie Z (2009) Chem Commun 2431–2445
66. Dang L, Lin Z, Marder TB (2009) Chem Commun 3987
67. Laitar DS, Müller P, Sadighi JP (2005) J Am Chem Soc 127:17196–17197

Chapter 7
Conclusion

This thesis describes (1) the synthesis and structural characterization of B-substituted nickel-1,2-o-carboryne complexes, (2) the reaction chemistry of Ni-1,2-o-carboryne with alkenes or/and alkynes, and (3) the formation of 1,3-o-carboryne and its reaction with alkynes catalyzed by transition metals.

Complexes (η^2-1,2-$C_2B_{10}R_n^1H_{10-n}$)Ni(PR_3^2)$_2$ (R^1 = I, n = 1, R_2 = Ph (**II-1**); R^1 = I, n = 2, R_2 = Ph (**II-2**); R^1 = Br, n = 1, R_2 = Me (**II-3**); R^1 = Ph, n = 1, R_2 = Me (**II-4**); R^1 = Ph, n = 1, R_2 = Ph (**II-5**)) were synthesized by salt elimination of phosphine ligated metal halide with dilithiocarboranes. The substituents on the carborane cage have significant effects on these complexes and **II-5** exhibits exceptional stability toward heat and moisture. B–H$\cdots$Ni interactions were observed in the IR spectra and solid-state structures of **II-4** and **II-5** due to the steric effect of the phenyl substituent.

In the reactivity study of (η^2-$C_2B_{10}H_{10}$)Ni(PPh$_3$)$_2$, we found alkenes can regioselectively react with nickelacarboryne in an insertion manner followed by a β-H elimination to afford alkenylcarboranes. Both aliphatic and aromatic, terminal and internal, cyclic and acyclic alkenes underwent the insertion reactions, and among them substituted styrenes gave the best results. The mechanism was supported by the D-substitued experiments. The β-H elimination cannot occur with some substrates such as methyl acrylate and 2-vinyl pyridine leading to the isolation of the thermodynamically stable inserted intermediates, nickelacyclopentanes.

In view of that nickelacyclopentane intermediates can react readily with alkynes to give dihydrobenzocarborane derivatives, a novel nickel-mediated three-component assembling reaction of 1,2-o-carboryne with alkenes and alkynes was developed. The formation of products can be rationalized by the sequential insertion of alkene and alkyne into the Ni–C(cage) and Ni–C(alkyl) bond, followed by reductive elimination. By the analogy between benzyne and 1,2-o-carboryne, nickel-catalyzed three-component cycloaddition reactions of arynes, activated alkenes, and alkynes have been achieved, leading to a series of substituted dihydronaphthalenes in moderate to very good isolated yields with excellent chemo- and regioselectivity.

Z. Qiu, *Late Transition Metal-Carboryne Complexes*, Springer Theses
DOI: 10.1007/978-3-642-24361-5_7, © Springer-Verlag Berlin Heidelberg 2012

1-Iodo-2-lithiocarborane, conveniently prepared in situ from the reaction of dilithiocarborane with 1 equiv of iodine, was used as the 1,2-o-carboryne precursor to develop a catalytic version of the reactions of 1,2-o-carboryne with alkynes. The isolated yields of the benzocarborane products were very comparable with those of stoichiometric reactions of Ni-1,2-o-carboryne with alkynes. The key intermediate, nickelacyclopentene [{[2-C(^{n}Bu)=C(o-C$_5$H$_4$N)-1,2-C$_2$B$_{10}$H$_{10}$]Ni}$_2$(μ_2-Cl)][Li(THF)$_4$] was isolated and structurally confirmed.

1,3-o-Carboryne, which can be regarded as a new boron nucleophile, can be formed by salt elimination of 3-iodo-1-lithio-o-carborane catalyzed by palladium(0). Due to high reaction temperatures, 1,3-o-carboryne cannot be stabilized by transition-metal, but can be trapped by alkynes to form [2+2+2] cycloaddition products. These studies introduce a direct and efficient route to the synthesis of 1,3-benzo-o-carborane derivatives and demonstrate the relative reactivity of M–C over M–B bond in 1,3-o-carboryne complexes toward alkynes.

Chapter 8
Experimental Section

General Procedures. All experiments were performed under an atmosphere of dry dinitrogen with the rigid exclusion of air and moisture using standard Schlenk or cannula techniques, or in a glovebox. All organic solvents (except CH_2Cl_2) were refluxed over sodium benzophenone ketyl for several days and freshly distilled prior to use. CH_2Cl_2 was refluxed over CaH_2 for several days and distilled immediately prior to use. All chemicals were purchased from either Aldrich or Acros Chemical Co. and used as received unless otherwise noted. $(Ph_3P)_2NiCl_2$ [1], $(Me_2Im)_2NiI_2$ [2], aryne precursors **IV-2b–e** [3, 4], 3-bromo-o-carborane [5], 3-iodo-o-carborane [6], 9-iodo-o-carborane [7], 9,12-diiodo-o-carborane [8], 1-phenyl-o-carborane [9] were prepared according to literature methods. Infrared spectra were obtained from KBr pellets prepared in the glovebox on a Perkin-Elmer 1,600 Fourier transform spectrometer. ^1H NMR spectra were recorded on either a Bruker DPX 300 spectrometer at 300 MHz or a Bruker DPX 400 spectrometer at 400 MHz. $^{13}C\{^1H\}$ NMR spectra were recorded on either a Bruker DPX 300 spectrometer at 75 MHz or a Bruker DPX 400 spectrometer at 100 MHz. $^{11}B\{^1H\}$ NMR spectra were recorded on either a Bruker DPX 300 spectrometer at 96 MHz or a Bruker DPX 400 spectrometer at 128 MHz. All chemical shifts were reported in δ units with references to the residual solvent resonances of the deuterated solvents for proton and carbon chemical shifts, to external $BF_3 \cdot OEt_2$ (0.00 ppm) for boron chemical shifts, and to external 85% H_3PO_4 (0.00 ppm) for phosphorous chemical shifts. Mass spectra were obtained on a Thermo Finnigan MAT 95 XL spectrometer. Elemental analyses were performed by the Shanghai Institute of Organic Chemistry, CAS, China.

Preparation of $(\eta^2$-9-I-1,2-$C_2B_{10}H_9)Ni(PPh_3)_2$ (II-1). A 1.6 M solution of n-BuLi in n-hexane (0.625 mL, 1.0 mmol) was slowly added to a stirring solution of 9-I-1,2-$C_2B_{10}H_{11}$ (135 mg, 0.5 mmol) in THF (10 mL) at 0 °C, and the mixture was stirred at room temperature for 1 h. The resulting 1,2-Li_2-9-I-1,2-$C_2B_{10}H_9$ suspension was then cooled to 0 °C, to which was added $(Ph_3P)_2NiCl_2$ (327 mg, 0.5 mmol). The reaction mixture was then stirred for 0.5 h at room temperature, giving a brown solution. After removal of the solvent, the deep brown residue was extracted with toluene (10 mL $\times$ 3). The brown filtrate was concentrated to 5 mL.

Z. Qiu, *Late Transition Metal-Carboryne Complexes*, Springer Theses, DOI: 10.1007/978-3-642-24361-5_8, © Springer-Verlag Berlin Heidelberg 2012

Complex **II-1**·0.5toluene was obtained as a yellow solid after it stood at room temperature for 2 days (247 mg, 55%). ^{1}H NMR (benzene-d_6): δ7.28 (m, 12H, C_6H_5), 6.86 (m, 18H, C_6H_5). ^{13}C{^{1}H} NMR (benzene-d_6): δ 133.7 (d, $^2J_{C-P}$ = 11.5 Hz), 131.9 (d, $^1J_{C-P}$ = 44.5 Hz), 130.6, 128.7 (d, $^3J_{C-P}$ = 6.1 Hz), the cage carbons are not observed. ^{11}B{^{1}H} NMR (benzene-d_6): δ −1.1 (3B), −13.9 (6B), −22.9 (1B). ^{31}P{^{1}H} NMR (benzene-d_6): δ 33.9. IR (KBr, cm^{-1}): ν_{BH} 2578 (vs). Anal. Calcd for $C_{83}H_{86}B_{20}Ni_2P_4I_2$ (**II-1** + 0.5toluene): C, 55.54; H, 4.83. Found: C, 55.68; H, 5.06.

Preparation of (η^2-9,12-I$_2$-1,2-C$_2$B$_{10}$H$_8$)Ni(PPh$_3$)$_2$ (II-2). A 1.6 M solution of n-BuLi in n-hexane (0.625 mL, 1.0 mmol) was slowly added to a stirring solution of 9,12-I$_2$-1,2-C$_2$B$_{10}$H$_{10}$ (198 mg, 0.5 mmol) in THF (10 mL) at 0 °C, and the mixture was stirred at room temperature for 1 h. The resulting 1,2-Li$_2$-9,12-I$_2$-1, 2-C$_2$B$_{10}$H$_8$ suspension was then cooled to 0 °C, to which was added (Ph$_3$P)$_2$NiCl$_2$ (327 mg, 0.5 mmol). The reaction mixture was then stirred for 0.5 h at room temperature, giving a brown solution. After removal of the solvent, the deep brown residue was extracted with CH$_2$Cl$_2$ (20 mL). The brown filtrate was concentrated to 3 mL. Complex **II-2** was obtained as yellow crystals after this solution stood at −30 °C overnight (352 mg, 72%). ^{1}H NMR (CD$_2$Cl$_2$): δ 7.40 (m, 6H, C_6H_5), 7.25 (m, 24H, C_6H_5). ^{13}C{^{1}H} NMR (CD$_2$Cl$_2$): δ 133.1 (dd, $^2J_{C-P}$ = 6.0 Hz), 130.3, 128.2 (dd, $^3J_{C-P}$ = 4.9 Hz), the cage carbons are not observed. ^{11}B{^{1}H} NMR (CD$_2$Cl$_2$): δ −0.5 (2B), −14.0 (6B), −21.6 (2B). ^{31}P{^{1}H} NMR (CD$_2$Cl$_2$): δ 32.4. IR (KBr, cm^{-1}): ν_{BH} 2592 (vs). Anal. Calcd for $C_{38}H_{38}B_{10}NiP_2I_2$ (**II-2**): C, 46.70; H, 3.92. Found: C, 47.19; H, 3.97.

Preparation of (η^2-3-Br-1,2-C$_2$B$_{10}$H$_9$)Ni(PMe$_3$)$_2$ (II-3). A 1.6 M solution of n-BuLi in n-hexane (0.625 mL, 1.0 mmol) was slowly added to a stirring solution of 3-Br-1,2-C$_2$B$_{10}$H$_{11}$ (112 mg, 0.5 mmol) in THF (10 mL) at 0 °C, and the mixture was stirred at room temperature for 1 h. The resulting 1,2-Li$_2$-3-Br-1, 2-C$_2$B$_{10}$H$_9$ suspension was then cooled to 0 °C, to which was added (Me$_3$P)$_2$NiCl$_2$ (141 mg, 0.5 mmol). The reaction mixture was then stirred for 0.5 h at room temperature, giving a brown solution. After removal of the solvent, the deep brown residue was extracted with toluene (10 mL × 3). The brown filtrate was concentrated to 5 mL. Complex **II-3** was obtained as yellow crystals after it stood at room temperature for 2 days (67 mg, 31%). ^{1}H NMR (benzene-d_6): δ 0.73 (m, 18H, CH$_3$). ^{13}C{^{1}H} NMR (benzene-d_6): δ 16.6 (m), the cage carbons are not observed. ^{11}B{^{1}H} NMR (benzene-d_6): δ −1.5 (1B), −8.4 (1B), −10.8 (2B), −11.6 (2B), −12.6 (2B), −14.1 (2B). ^{31}P{^{1}H} NMR (benzene-d_6): δ −9.4. IR (KBr, cm^{-1}): ν_{BH} 2551 (vs). Anal. Calcd for $C_8H_{27}B_{10}BrNiP_2$ (**II-3**): C, 22.25; H, 6.30. Found: C, 22.61; H, 6.18.

Preparation of (η^2-3-C$_6$H$_5$-1,2-C$_2$B$_{10}$H$_9$)Ni(PMe$_3$)$_2$ (II-4). A 1.6 M solution of n-BuLi in n-hexane (0.625 mL, 1.0 mmol) was slowly added to a stirring solution of 3-C$_6$H$_5$-1,2-C$_2$B$_{10}$H$_{11}$ (110 mg, 0.5 mmol) in THF (10 mL) at 0 °C, and the mixture was stirred at room temperature for 1 h. The resulting 1,2-Li$_2$-3-C$_6$H$_5$-1,2-C$_2$B$_{10}$H$_9$ suspension was then cooled to 0 °C, to which was added (Me$_3$P)$_2$NiCl$_2$ (141 mg, 0.5 mmol). The reaction mixture was then stirred for 0.5 h at room temperature, giving a brown solution. After removal of the solvent, the deep brown residue was extracted with toluene (10 mL × 3). The brown filtrate

8 Experimental Section

was concentrated to 6–7 mL. Complex **II-4** was obtained as yellow crystals after it stood at room temperature for 3 days (90 mg, 42%). ^{1}H NMR (CD$_2$Cl$_2$): δ 7.82 (m, 2H, C$_6$H$_5$), 7.30 (m, 3H, C$_6$H$_5$), 1.16 (m, 18H, CH$_3$). ^{13}C{^{1}H} NMR (CD$_2$Cl$_2$): δ 133.6, 127.6, 126.9 (C$_6$H$_5$), 16.3 (m) (CH$_3$), the cage carbons are not observed. ^{11}B{^{1}H} NMR (CD$_2$Cl$_2$): δ −2.8 (1B), −3.8 (1B), −8.2 (1B), −12.2 (2B), −15.2 (5B). ^{31}P{^{1}H} NMR (CD$_2$Cl$_2$): δ −9.1. IR (KBr, cm^{-1}): ν_{BH} 2550, 2531 (vs). Anal. Calcd for C$_{14}$H$_{32}$B$_{10}$NiP$_2$ (**II-4**): C, 39.18; H, 7.52. Found: C, 38.96; H, 7.71.

Preparation of (η^2-3-C$_6$H$_5$-1,2-C$_2$B$_{10}$H$_9$)Ni(PPh$_3$)$_2$ (II-5). A 1.6 M solution of n-BuLi in n-hexane (0.625 mL, 1.0 mmol) was slowly added to a stirring solution of 3-C$_6$H$_5$-1,2-C$_2$B$_{10}$H$_{11}$ (110 mg, 0.5 mmol) in THF (10 mL) at 0 °C, and the mixture was stirred at room temperature for 1 h. The resulting 1,2-Li$_2$-3-C$_6$H$_5$-1,2-C$_2$B$_{10}$H$_9$ suspension was then cooled to 0 °C, to which was added (Ph$_3$P)$_2$NiCl$_2$ (327 mg, 0.5 mmol). The reaction mixture was then stirred for 0.5 h at room temperature, giving a brown solution. After removal of the solvent, the deep brown residue was extracted with toluene (10 mL × 3). The brown filtrate was concentrated to 10 mL. Complex **II-5** was obtained as orange crystals after it stood at room temperature for 3 days (305 mg, 76%). ^{1}H NMR (CD$_2$Cl$_2$): δ 7.70 (d, J = 7.2 Hz, 2H, BC$_6$H$_5$), 7.45 (t, J = 7.2 Hz, 1H, BC$_6$H$_5$), 7.33 (m, 6H, PC$_6$H$_5$), 7.18 (m, 14H, BC$_6$H$_5$ & PC$_6$H$_5$), 7.06 (m, 12H, PC$_6$H$_5$). ^{13}C{^{1}H} NMR (CD$_2$Cl$_2$): δ 135.3 (s, BC$_6$H$_5$), 134.0 (m, PC$_6$H$_5$), 131.9 (d, $^1J_{C-P}$ = 44.2 Hz, PC$_6$H$_5$), 130.5 (s, PC$_6$H$_5$), 128.6 (m, PC$_6$H$_5$), 128.1 (BC$_6$H$_5$), the cage carbons are not observed. ^{11}B{^{1}H} NMR (CD$_2$Cl$_2$): δ −1.7 (2B), −7.7 (1B), −11.9 (3B), −13.4 (4B). ^{31}P{^{1}H} NMR (CD$_2$Cl$_2$): δ 29.6. IR (KBr, cm^{-1}): ν_{BH} 2557, 2512 (vs). Anal. Calcd for C$_{44}$H$_{44}$B$_{10}$NiP$_2$ (**II-5**): C, 65.93; H, 5.53. Found: C, 66.12; H, 5.54.

General Procedure for Nickel-Mediated Cycloaddition Reaction of 1,2-o-Carboryne with Alkynes. To a THF solution (5 mL) of Li$_2$C$_2$B$_{10}$H$_{10}$ (1.0 mmol), prepared in situ from the reaction of n-BuLi (2.0 mmol) with o-carborane (1.0 mmol), was added (PPh$_3$)$_2$NiCl$_2$ (1.0 mmol). The reaction mixture was stirred at room temperature for 0.5 h to give the Ni-1,2-o-carboryne complex (η^2-C$_2$B$_{10}$H$_{10}$)Ni(PPh$_3$)$_2$ [10]. Alkene (2.0 mmol) was then added and the reaction vessel was closed and heated at 90 °C overnight. After removal of the precipitate, the resulting solution was concentrated to dryness in vacuo. The residue was subject to column chromatography on silica gel (40–230 mesh) to give the coupling product.

$trans$-1-(HC=CHPh)-1,2-C$_2$B$_{10}$H$_{11}$ (III-3a). Yield: 82%. Colorless oil. ^{1}H NMR (400 MHz, CDCl$_3$): δ 7.34 (s, 5H) (Ph), 6.85 (d, J = 16.0 Hz, 1H), 6.28 (d, J = 16.0 Hz, 1H) (olefinic), 3.72 (s, 1H) (cage CH). ^{13}C{^{1}H} NMR (100 MHz, CDCl$_3$): δ 137.7, 129.5, 128.9, 126.9, 122.5 (olefinic and Ph), 60.9 (cage C), another cage carbon was not observed. ^{11}B{^{1}H} NMR (128 MHz, CDCl$_3$): δ −1.9 (1B), −4.5 (1B), −8.6 (2B), −10.2 (4B), −12.0 (2B). ^{1}H NMR (300 MHz, benzene-d_6): δ 6.99 (m, 3H), 6.88 (m, 2H) (Ph), 6.48 (d, J = 15.9 Hz, 1H), 5.76 (d, J = 15.9 Hz, 1H) (olefinic), 2.47 (s, 1H) (cage CH). ^{11}B{^{1}H} NMR (96 MHz, benzene-d_6): δ −2.2 (1B), −5.0 (1B), −9.4 (2B), −11.3 (2B), −12.2 (2B), −13.2 (2B). HRMS: m/z calcd for C$_{10}$H$_{18}^{11}$B$_8^{10}$B$_2^+$: 246.2406. Found: 246.2407.

104   8 Experimental Section

trans-1-[DC=CD(Ph)]-2-D-1,2-$C_2B_{10}H_{11}$ ([D_3]-III-3a). Yield: 80%. Colorless oil. ^1H NMR (400 MHz, benzene-d_6): δ 6.99 (m, 3H), 6.87 (m, 2H) (Ph). ^2H NMR (61 MHz, benzene): δ 6.47 (1^2H), 5.76 (1^2H) (olefinic), 2.42 (1^2H) (cage C^2H). HRMS: m/z calcd for $C_{10}H_{15}^2H_3^{11}B_8^{10}B_2^+$: 249.2595. Found: 249.2588.

trans-1-{HC=CH[($2'$-CH_3)C_6H_4]}-1,2-$C_2B_{10}H_{11}$ (III-3b). Yield: 85%. White solid. ^1H NMR (300 MHz, CDCl$_3$): δ 7.23 (d, J = 8.1 Hz, 2H), 7.15 (d, J = 8.1 Hz, 2H) (Ph), 6.81 (d, J = 15.9 Hz, 1H), 6.23 (d, J = 15.9 Hz, 1H) (olefinic), 3.71 (s, 1H) (cage CH), 2.35 (s, 3H) (CH_3). $^{13}C\{^1H\}$ NMR (100 MHz, CDCl$_3$): δ 139.7, 137.6, 131.4, 129.6, 126.9, 121.4 (olefinic and Ph), 74.4, 61.0 (cage C), 21.3 (CH_3). $^{11}B\{^1H\}$ NMR (128 MHz, CDCl$_3$): δ −2.7 (1B), −5.8 (1B), −9.9 (2B), −11.7 (2B), −12.3 (2B), −13.6 (2B). HRMS: m/z calcd for $C_{11}H_{20}^{11}B_8^{10}B_2^+$: 260.2563. Found: 260.2561.

trans-1-{HC=CH[($4'$-CF_3)C_6H_4]}-1,2-$C_2B_{10}H_{11}$ (III-3c). Yield 80%. White solid. ^1H NMR (300 MHz, CDCl$_3$): δ 7.61 (d, J = 8.1 Hz, 2H), 7.45 (d, J = 8.1 Hz, 2H) (Ph), 6.88 (d, J = 15.9 Hz, 1H), 6.36 (d, J = 15.9 Hz, 1H) (olefinic), 3.74 (s, 1H) (cage CH). $^{13}C\{^1H\}$ NMR (100 MHz, CDCl$_3$): δ 137.5, 136.2, 127.2, 125.9, 125.2 (CF_3, olefinic and Ph), 73.3, 60.7 (cage C). $^{11}B\{^1H\}$ NMR (128 MHz, CDCl$_3$): δ −2.5 (1B), −5.3 (1B), −9.7 (2B), −11.7 (2B), −12.3 (2B), −13.4 (2B). HRMS: m/z calcd for $C_{11}H_{17}^{11}B_8^{10}B_2F_3^+$: 314.2280. Found: 314.2275.

trans-1-{HC=CH[($3'$-CF_3)C_6H_4]}-1,2-$C_2B_{10}H_{11}$ (III-3d). Yield 73%. White solid. ^1H NMR (300 MHz, CDCl$_3$): δ 7.52 (m, 4H) (Ph), 6.90 (d, J = 15.9 Hz, 1H), 6.35 (d, J = 15.9 Hz, 1H) (olefinic), 3.74 (s, 1H) (cage CH). $^{13}C\{^1H\}$ NMR (100 MHz, CDCl$_3$): δ 136.2, 134.9, 130.1, 129.5, 126.0, 124.5, 123.6, 123.5 (CF_3, olefinic and Ph), 73.4, 60.7 (cage C). $^{11}B\{^1H\}$ NMR (128 MHz, CDCl$_3$): δ −2.0 (1B), −4.8 (1B), −9.2 (2B), −11.2 (2B), −11.9 (2B), −12.9 (2B). HRMS: m/z calcd for $C_{11}H_{17}^{11}B_8^{10}B_2F_3^+$: 314.2280. Found: 314.2276.

trans-1-{HC=CH[$3',4',5'$-(OMe)$_3C_6H_2$]}-1,2-$C_2B_{10}H_{11}$ (III-3e). Yield 76%. White solid. ^1H NMR (300 MHz, CDCl$_3$): δ 6.78 (d, J = 15.6 Hz, 1H) (olefinic), 6.54 (s, 2H) (Ph), 6.17 (d, J = 15.6 Hz, 1H) (olefinic), 3.88 (s, 6H), 3.85 (s, 3H) (OCH_3), 3.72 (s, 1H) (cage CH). $^{13}C\{^1H\}$ NMR (100 MHz, CDCl$_3$): δ 153.5, 139.4, 137.7, 129.6, 121.7, 104.2 (olefinic and Ph), 74.1(cage C), 61.0, 56.2 (OCH_3), another cage carbon was not observed. $^{11}B\{^1H\}$ NMR (128 MHz, CDCl$_3$): δ −2.5 (1B), −5.6 (1B), −9.8 (2B), −11.6 (2B), −12.3 (2B), −13.5 (2B). HRMS: m/z calcd for $C_{13}H_{24}^{11}B_8^{10}B_2^+$: 336.2723. Found: 336.2718.

1-[H_2CC(Ph)=CH_2]-1,2-$C_2B_{10}H_{11}$ (III-4f). Yield: 59%. Colorless oil. ^1H NMR (300 MHz, CDCl$_3$): δ 7.36 (m, 5H) (Ph), 5.48 (s, 1H), 5.20 (s, 1H) (olefinic), 3.49 (s, 2H) (CH_2), 3.37 (s, 1H) (cage CH). $^{13}C\{^1H\}$ NMR (100 MHz, CDCl$_3$): δ 142.8, 139.1, 128.9, 128.6, 126.2, 119.7 (olefinic and Ph), 73.8, 58.8 (cage C), 42.5 (CH_2). $^{11}B\{^1H\}$ NMR (128 MHz, CDCl$_3$): δ −2.5 (1B), −5.9 (1B), −9.9 (2B), −11.1 (2B), −12.7 (2B), −13.4 (2B). HRMS: m/z calcd for $C_{11}H_{20}^{11}B_8^{10}B_2^+$: 260.2563. Found: 260.2563.

1-[HC=C(Ph)$_2$]-1,2-$C_2B_{10}H_{11}$ (III-3g). Yield: 46%. Colorless oil. ^1H NMR (400 MHz, CDCl$_3$): δ 7.46 (m, 3H), 7.28 (m, 3H), 7.17 (m, 4H) (Ph), 6.27 (s, 1H) (olefinic), 3.00 (s, 1H) (cage CH). $^{13}C\{^1H\}$ NMR (100 MHz, CDCl$_3$): δ 145.2,

8 Experimental Section

140.0, 136.0, 129.1, 129.0, 128.9, 128.8, 128.5, 126.9, 122.0 (olefinic and Ph), 73.2, 57.5 (cage C). $^{11}B\{^1H\}$ NMR (128 MHz, CDCl$_3$): δ −2.5 (1B), −4.2 (1B), −9.8 (4B), −10.4 (2B), −13.1 (2B). HRMS: m/z calcd for $C_{16}H_{22}^{11}B_8^{10}B_2{}^+$: 322.2719. Found: 322.2716.

trans-1-[HC=CH(SiMe$_3$)]-1,2-C$_2$B$_{10}$H$_{11}$ (III-3h). Yield: 46%. Colorless oil. ^1H NMR (400 MHz, CDCl$_3$): δ 6.24 (d, $J = 18.4$ Hz, 1H), 6.01 (d, $J = 18.4$ Hz, 1H) (olefinic), 3.65 (s, 1H) (cage CH), 0.08 (s, 9H) (CH_3). $^{13}C\{^1H\}$ NMR (100 MHz, CDCl$_3$): δ 139.4, 137.2 (olefinic), 59.8 (cage C), −1.8 (CH_3), another cage carbon was not observed. $^{11}B\{^1H\}$ NMR (96 MHz, CDCl$_3$): δ −2.3 (1B), −5.0 (1B), −9.2 (2B), −11.5 (4B), −13.0 (2B). HRMS: m/z calcd for $C_7H_{22}^{11}B_8^{10}B_2Si^+$: 242.2488. Found: 242.2483.

1-[H$_2$CC=CH(CH$_2$)$_3$CH$_2$]-1,2-C$_2$B$_{10}$H$_{11}$ (III-4i). Yield: 77%. Colorless oil. ^1H NMR (400 MHz, benzene-d_6): δ 5.07 (s, 1H) (olefinic), 2.62 (s, 1H) (cage CH), 2.20 (s, 2H) (CB–CH_2), 1.72 (m, 2H), 1.62 (m, 2H), 1.30 (m, 4H) (CH_2). $^{13}C\{^1H\}$ NMR (benzene-d_6): δ 132.8 (olefinic), 74.9, 60.6 (cage C), 45.9 (acyclic CH_2), 29.3, 25.3, 22.7, 21.7 (cyclic CH_2). $^{11}B\{^1H\}$ NMR (128 MHz, benzene-d_6): δ −3.2 (1B), −6.5 (1B), −9.8 (2B), −11.6 (2B), −13.7 (4B). HRMS: m/z calcd for $C_9H_{22}^{11}B_8^{10}B_2{}^+$: 238.2719. Found: 238.2718.

cis-/trans-1-[H$_2$CCH=CH(^{n}Pr)]-1,2-C$_2$B$_{10}$H$_{11}$ (III-4j). Yield: 74%. Colorless oil. ^1H NMR (300 MHz, CDCl$_3$): δ 5.64 (m, 1H), 5.53 (m, 1H), 5.31 (m, 2H) (olefinic), 3.60 (s, 1H), 3.56 (s, 1H) (cage CH), 2.98 (d, $J = 7.8$ Hz, 2H), 2.88 (d, $J = 7.5$ Hz, 2H), 2.00 (m, 4H), 1.40 (m, 4H) (CH_2), 0.90 (t, $J = 7.5$ Hz, 3H), 0.89 (t, $J = 7.5$ Hz, 3H) (CH_3). $^{13}C\{^1H\}$ NMR (100 MHz, CDCl$_3$): δ 137.5, 135.7, 122.9, 122.1 (olefinic), 74.5, 59.4 (cage C), 40.7, 34.8, 34.3, 29.3, 22.5, 22.2 (CH_2), 13.7, 13.6 (CH_3). $^{11}B\{^1H\}$ NMR (128 MHz, CDCl$_3$): δ −1.3 (1B), −4.7 (1B), −8.2 (2B), −10.2 (2B), −12.4 (4B). HRMS: m/z calcd for $[C_7H_{22}^{11}B_8^{10}B_2−2H]^+$: 224.2563. Found: 224.2552.

1-[HCC=CH(CH$_2$)$_2$CH$_2$]-1,2-C$_2$B$_{10}$H$_{11}$ (III-4k). Yield: 67%. Colorless oil. ^1H NMR (400 MHz, CDCl$_3$): δ 5.86 (m, 1H), 5.57 (d, $J = 10.4$ Hz, 1H) (olefinic), 3.70 (s, 1H) (cage CH), 2.97 (m, 1H) (CH), 1.98 (m, 3H), 1.79 (m, 1H), 1.54 (m, 2H) (CH_2). $^{13}C\{^1H\}$ NMR (100 MHz, CDCl$_3$): δ 131.3, 125.8 (olefinic), 80.0, 59.7 (cage C), 40.8, 29.6, 24.4, 21.1 (CH_2). $^{11}B\{^1H\}$ NMR (128 MHz, CDCl$_3$): δ −1.8 (1B), −3.8 (1B), −8.3 (2B), −10.7 (2B), −12.8 (4B). These data are identical with those reported in the literature [11].

1-bicyclo[2.2.1]hept-2-yl-1,2-carborane (III-5l). Yield: 60%. Colorless oil. ^1H NMR (400 MHz, benzene-d_6): δ 2.43 (s, 1H) (cage CH), 1.85 (m, 2H), 1.46 (m, 1H) (CH), 1.16 (m, 2H), 1.04 (m, 2H) 0.70 (m, 4H) (CH_2). $^{13}C\{^1H\}$ NMR (100 MHz, benzene-d_6): δ 81.3, 62.5 (cage C), 49.0, 43.8, 39.6 (CH), 36.6, 35,7, 30.2, 28.4 (CH_2). $^{11}B\{^1H\}$ NMR (128 MHz, benzene-d_6): δ −3.1 (1B), −5.2 (1B), −9.5 (2B), −12.2 (2B), −13.6 (4B). ^1H NMR (300 MHz, CDCl$_3$): δ 3.57 (s, 1H) (cage CH), 2.31 (m, 2H), 2.12 (m, 1H) (CH), 1.61 (m, 2H), 1.53 (m, 1H), 1.35 (m, 1H), 1.13 (m, 4H) (CH_2). $^{11}B\{^1H\}$ NMR (96 MHz, CDCl$_3$): δ −3.4 (2B),

−5.6 (1B), −9.8 (2B), −12.3 (2B), −12.6 (2B), −13.9 (2B). HRMS: m/z calcd for $C_9H_{22}^{11}B_8^{10}B_2^+$: 238.2719. Found: 238.2710.

1-(1H-inden-2-yl)-1,2-carborane (III-3m). Yield: 31%. White solid. ^{1}H NMR (400 MHz, CDCl$_3$): δ 7.35 (m, 4H), 6.98 (s, 1H) (aromatic), 3.85 (s, 1H) (cage C*H*), 3.50 (s, 2H) (C*H*$_2$). ^{13}C{^{1}H} NMR (100 MHz, CDCl$_3$): δ 142.5, 142.3, 139.9., 133.9, 127.2, 126.6, 123.8, 122.0 (aromatic and olefinic), 73.6, 61.4 (cage *C*), 42.0 (C*H*$_2$). ^{11}B{^{1}H} NMR (128 MHz, CDCl$_3$): δ −2.3 (1B), −4.8 (1B), −9.2 (2B), −11.0 (2B), −11.4 (2B), −12.9 (2B). These data are identical with those reported in the literature [12].

1-(2,3-dihydro-1H-inden-2-yl)-1,2-carborane (III-5m). Yield: 27%. Colorless oil. ^{1}H NMR (300 MHz, CDCl$_3$): δ 7.17 (s, 4H) (aromatic), 3.68 (s, 1H) (cage C*H*), 3.03 (m, 5H) (C*H* and C*H*$_2$). ^{13}C{^{1}H} NMR (100 MHz, CDCl$_3$): δ 140.4, 127.28, 124.3 (aromatic), 61.2 (cage *C*), 47.3 (*C*H), 39.9 (C*H*$_2$). ^{11}B{^{1}H} NMR (128 MHz, CDCl$_3$): δ −2.1 (1B), −4.4 (1B), −8.6 (2B), −10.9 (2B), −12.1 (2B), −12.5 (2B). HRMS: m/z calcd for $C_{11}H_{20}^{11}B_8^{10}B_2^+$: 260.2563. Found: 260.2561.

***trans*-1-[HC=CH(O^{n}Bu)]-1,2-C$_2$B$_{10}$H$_{11}$ (III-3n).** Yield: 18%. Colorless oil. ^{1}H NMR (400 MHz, CDCl$_3$): δ 6.81 (d, J = 12.4 Hz, 1H), 5.07 (d, J = 12.4 Hz, 1H) (olefinic), 3.66 (t, J = 6.4 Hz, 2H) (OC*H*$_2$), 3.54 (s, 1H) (cage C*H*), 1.61 (m, 2H), 1.38 (m, 2H) (C*H*$_2$), 0.93 (t, J = 7.6 Hz, 3H) (C*H*$_3$). ^{13}C{^{1}H} NMR (100 MHz, CDCl$_3$): δ 155.2, 100.2 (olefinic), 70.4 (cage *C*), 62.4 (OC*H*$_2$), 31.0, 19.0 (C*H*$_2$), 13.7 (C*H*$_3$), another cage carbon was not observed.. ^{11}B{^{1}H} NMR (96 MHz, CDCl$_3$): δ −1.3 (1B), −5.5 (1B), −9.5 (2B), −10.3 (2B), −11.4 (2B), −12.5 (2B). HRMS: m/z calcd for $C_8H_{22}O^{11}B_8^{10}B_2^+$: 242.2688. Found: 242.2682.

1-[HC(Me)(O^{n}Bu)]-1,2-C$_2$B$_{10}$H$_{11}$ (III-5n). Yield: 12%. Colorless oil. ^{1}H NMR (400 MHz, CDCl$_3$): δ 4.08 (s, 1H) (cage C*H*), 3.86 (q, J = 6.8 Hz, 1H) (OC*H*), 3.56 (m, 1H) (OCH*H*), 3.31 (m, 1H) (OCH*H*), 1.52 (m, 2H), 1.33 (m, 5H) (C*H*$_2$ and C*H*$_3$), 0.91 (t, J = 6.8 Hz, 3H) (C*H*$_3$). ^{13}C{^{1}H} NMR (100 MHz, CDCl$_3$): δ 78.2, 58.3 (cage *C*), 75.5 (O*C*H), 70.3 (O*C*H$_2$), 31.7, 19.7, 19.3 (C*H*$_2$), 13.8 (C*H*$_3$). ^{11}B{^{1}H} NMR (96 MHz, CDCl$_3$): δ −3.9 (1B), −4.7 (1B), −9.3 (3B), −12.1 (3B), −13.4 (1B), −14.3 (1B). HRMS: m/z calcd for $C_8H_{24}O^{11}B_8^{10}B_2^+$: 244.2825. Found: 244.2823.

1-[HCC=CH(CH$_2$)$_2$O]-1,2-C$_2$B$_{10}$H$_{11}$ (III-4o): Yield 15%. Colorless oil. ^{1}H NMR (400 MHz, CDCl$_3$): δ 6.04 (m, 1H) (olefinic), 5.72 (d, J = 6.9 Hz, 1H) (olefinic), 4.54 (s, 1H) (OC*H*), 4.09 (s, 1H) (cage C*H*), 3.98 (ddd, J = 1.2, 4.2, 8.4 Hz, 1H) (OCH*H*), 3.65 (dt, J = 2.7, 8.4 Hz, 1H) (OCH*H*), 2.28 (m, 1H) (CH*H*), 1.93 (ddd, J = 1.2, 2.7, 13.2 Hz, 1H) (CH*H*). ^{13}C{^{1}H} NMR (100 MHz, CDCl$_3$): δ 127.5, 126.0 (olefinic), 73.3 (O*C*H), 64.3 (cage *C*), 58.3 (O*C*H$_2$), 25.3 (C*H*$_2$), another cage carbon was not observed. ^{11}B{^{1}H} NMR (128 MHz, CDCl$_3$): δ −3.4 (1B), −4.4 (1B), −9.4 (2B), −12.1 (3B), −13.8 (3B). HRMS: m/z calcd for $C_7H_{18}O^{11}B_8^{10}B_2^+$: 226.2355. Found: 226.2357.

1-[CH$_2$CH$_2$(CO$_2$Me)]-1,2-C$_2$B$_{10}$H$_{11}$ (III-5p). Yield: 62%. White solid. ^{1}H NMR (400 MHz, CDCl$_3$): δ 3.70 (s, 4H) (OC*H*$_3$ and cage C*H*), 2.55 (m, 4H) (C*H*$_2$). ^{1}H NMR (400 MHz, benzene-d_6): δ 3.21 (s, 3H) (OC*H*$_3$), 2.58 (s, 1H)

8 Experimental Section

(cage CH), 1.88 (s, 4H) (CH_2). $^{13}C\{^1H\}$ NMR (100 MHz, CDCl$_3$): δ 171.8 (C=O), 73.9, 61.5 (cage C), 52.2 (OCH_3), 33.2, 32.7 (CH_2). $^{11}B\{^1H\}$ NMR (128 MHz, CDCl$_3$): δ −2.3 (1B), −5.7 (1B), −9.6 (2B), −11.8 (2B), −12.3 (2B), −13.0 (2B). HRMS: m/z Calcd for $C_6H_{18}O_2^{11}B_8^{10}B_2{}^+$: 230.2304. Found: 230.2302.

1-[CH$_2$CH(D)(CO$_2$Me)]-2-D-1,2-C$_2$B$_{10}$H$_{11}$ ([D$_2$]-III-5p). Methyl acrylate (172 mg, 2.0 mmol) was added to the THF suspension of Ni-1,2-o-carboryne (1.0 mmol) prepared in situ from dilithiocarborane and NiCl$_2$(PPh$_3$)$_2$ [10], and the mixture was heated at 90 °C overnight. D$_2$O (2 mL) was added and the reaction mixture was heated at 60 °C for 3 h. After removal of the solvent under vacuum, the oily residue was purified by column chromatography on silica gel (230–400 mesh) using hexane/ether (v/v = 15/1) as eluent to afford **[D$_2$]-III-5p** as a white solid (125 mg, 54%). ^1H NMR (300 MHz, benzene-d_6): δ 3.19 (s, 3H) (OCH_3), 1.88 (s, 3H) (CHD and CH_2). ^2H NMR (61 MHz, benzene-d_6): δ 2.63 (1^{2}H) (cage C^2H), 1.87 (1^{2}H) (CH^2H). HRMS: m/z Calcd for $C_6H_{16}^2H_2O_2^{11}B_8^{10}B_2{}^+$: 232.2430. Found: 232.2430.

1-[CH$_2$CH(CO$_2$Me)CH$_2$CH$_2$CO$_2$Me]-1,2-C$_2$B$_{10}$H$_{11}$ (III-6p). Yield: 14%. Colorless oil. ^1H NMR (400 MHz, CDCl$_3$): δ 3.29 (s, 3H), 3.18 (s, 3H) (OCH_3), 2.98 (s, 1H) (cage CH), 2.40 (m, 2H), 1.90 (m, 2H), 1.63 (m, 1H), 1.39(m, 2H) (CH & CH_2). $^{13}C\{^1H\}$ NMR (100 MHz, CDCl$_3$): δ 174.4, 172.6 (C=O), 73.7, 61.2 (cage C), 52.4, 51.9 (OCH_3), 44.4 (CH), 39.2, 31.0, 28.2 (CH_2). $^{11}B\{^1H\}$ NMR (128 MHz, CDCl$_3$): δ −2.8 (1B), −6.1 (1B), −10.3 (3B), −13.6 (5B). HRMS: m/z Calcd for $C_{10}H_{24}O_4^{11}B_8^{10}B_2$ [M-H]$^+$: 315.2594. Found: 315.2594.

1-[CH$_2$CH$_2$(o-C$_5$H$_4$N)]-1,2-C$_2$B$_{10}$H$_{11}$ (III-5q). Yield: 59%. Colorless oil. ^1H NMR (400 MHz, CDCl$_3$): δ 8.50 (d, J = 4.4 Hz, 1H), 7.65 (dt, J = 2.0, 7.6 Hz, 1H), 7.18 (m, 2H) (Py), 3.80 (s, 1H) (cage CH), 2.98 (t, J = 8.0 Hz, 2H), 2.74 (t, J = 8.0 Hz, 2H) (CH_2). $^{13}C\{^1H\}$ NMR (100 MHz, CDCl$_3$): δ 158.2, 149.1, 137.0, 123.2, 122.0 (Py), 74.8, 61.5 (cage C), 36.9, 36.8 (CH_2). $^{11}B\{^1H\}$ NMR (128 MHz, CDCl$_3$): δ −2.6 (1B), −6.0 (1B), −9.6 (2B), −11.7 (2B), −12.4 (2B), −13.3 (2B). HRMS: m/z Calcd for $C_9H_{17}N^{11}B_8^{10}B_2{}^+$: 247.2364. Found: 247.2373.

1,2-[CH$_2$CH$_2$(o-C$_5$H$_4$N)]$_2$-1,2-C$_2$B$_{10}$H$_{10}$ (III-7q). Yield: 16%. Colorless crystals. ^1H NMR (400 MHz, CDCl$_3$): δ 8.51 (d, J = 4.4 Hz, 2H), 7.67 (m, 3H), 7.21 (m, 4H) (Py), 3.07 (m, 4H), 2.77 (m, 4H) (CH_2). $^{13}C\{^1H\}$ NMR (100 MHz, CDCl$_3$): δ 158.6, 149.3, 136.9, 123.2, 121.9 (Py), 79.5 (cage C), 37.5, 34.1 (CH_2). $^{11}B\{^1H\}$ NMR (128 MHz, CDCl$_3$): δ −4.8 (2B), −10.5 (8B). HRMS: m/z Calcd for $C_{16}H_{26}N_2^{11}B_8^{10}B_2{}^+$: 354.3099. Found: 354.3098.

***trans*-1-[CH=CH(o-C$_5$H$_4$N)]-2-[CH$_2$CH$_2$(o-C$_5$H$_4$N)]-1,2-C$_2$B$_{10}$H$_{10}$(III-8q).** Yield: 10%. Colorless crystals. ^1H NMR (400 MHz, CDCl$_3$): δ 8.56 (d, J = 4.8 Hz, 1H), 8.47 (d, J = 5.6 Hz, 1H), 7.70 (t, J = 7.6 Hz, 1H), 6.40 (t, J = 7.6 Hz, 1H), 7.33 (d, J = 7.6 Hz, 1H), 7.24 (dd, J = 4.8, 7.6 Hz, 1H), 7.16 (m, 2H) (Py), 3.06 (m, 2H), 2.70 (m, 2H) (CH_2). $^{13}C\{^1H\}$ NMR (100 MHz, CDCl$_3$): δ 158.6, 152.4, 149.9, 149.2, 140.0, 136.9, 136.7, 124.8, 123.9, 123.7, 123.1, 121.7 (Py & olefinic C), 80.0, 79.0 (cage C), 37.6, 34.5 (CH_2). $^{11}B\{^1H\}$ NMR (128 MHz, CDCl$_3$): δ −4.3 (2B), −10.4 (8B). HRMS: m/z Calcd for $C_{16}H_{24}N_2^{11}B_8^{10}B_2{}^+$: 352.2943. Found: 352.2931.

Preparation of Nickelacyclopentane (III-9p, III-9q). To a THF solution (5 mL) of $Li_2C_2B_{10}H_{10}$ (1.5 mmol), prepared in situ from the reaction of ^nBuLi (3.0 mmol) with o-carborane (1.5 mmol), was added $NiCl_2(PPh_3)_2$ (1.5 mmol). The reaction mixture was stirred at room temperature for 0.5 h to give the Ni-1, 2-o-carboryne complex [10]. After the addition of alkene (1.8 mmol), the reaction vessel was closed and heated at 90 °C overnight. Removal of the solvent gave a red residue which was extracted with ether (10 mL) twice. The combined ether solution was concentrated to dryness and washed with hexane (50 mL) three times. Removal of the solvent afforded **III-9** as a red solid. Recrystallization of **III-9p** from THF/hexane gave red microcrystals. Recrystallization of **III-9q** from DME (1,2-dimethoxyethane) afforded X-ray-quality crystals.

[{[2-CH$_2$CH(o-C$_5$H$_4$N)-1,2-C$_2$B$_{10}$H$_{10}$]Ni}$_3$(μ_3-Cl)][Li(DME)$_3$] (III-9q). Yield: 32%. Red crystals. ^{1}H NMR (400 MHz, CD$_2$Cl$_2$): δ 8.21 (m, 1H), 7.58 (d, $J = 8.2$ Hz, 1H), 7.42 (m, 1H), 6.70 (m, 1H) (Py), 6.08 (dd, $J = 6.1, 10.3$ Hz, 1H) (Ni–CH), 3.65 (s, 4H), 3.47 (s, 6H) (DME), 2.65 (dd, $J = 10.3, 14.6$ Hz, 1H) (CHH), 2.40 (dd, $J = 6.1, 14.6$ Hz, 1H) (CHH). ^{13}C{^{1}H} NMR (100 MHz, CD$_2$Cl$_2$): δ 172.1, 149.0, 135.6, 121.3, 115.6 (Py), 91.3, 76.5 (cage C), 70.5, 59.0 (DME), 43.8 (CH$_2$), 43.2 (Ni–CH). ^{11}B{^{1}H} NMR (96 MHz, CD$_2$Cl$_2$): δ -5.9 (2B), -8.7 (4B), -10.8 (4B). IR (KBr, cm^{-1}):ν 2976 (s), 2876 (s), 2575 (vs), 1598 (s), 1471 (s), 1045 (s), 1017 (s), 889 (m), 734 (m). Anal. Calcd for $C_{39}H_{81}B_{30}ClLiN_3Ni_3O_6$ (**III-9q**): C, 38.06; H, 6.63; N, 3.41. Found: C, 37.80; H, 6.47; N, 3.13.

[2-CH$_2$CH(CO$_2$Me)-1,2-C$_2$B$_{10}$H$_{10}$]Ni(PPh$_3$) (III-9r). Yield: 39%. Red solid. ^{1}H NMR (400 MHz, benzene-d_6): δ 7.60 (m, 6H), 6.99 (m, 9H) (PPh$_3$), 4.12 (dd, $J = 7.6, 16.0$ Hz, 1H) (CHH), 2.89 (s, 3H) (OCH_3), 2.67 (dd, $J = 9.2, 16.0$ Hz, 1H) (CHH), 1.74 (dd, $J = 7.6, 9.2$ Hz, 1H) (Ni–CH). ^{1}H NMR (400 MHz, pyridine-d_5): δ 7.42 (m, 6H), 7.32 (m, 9H) (PPh$_3$), 3.61 (m, 1H) (CHH), 3.01 (m, 1H) (CHH), 2.78 (s, 3H) (OCH_3), 2.00 (m, 1H) (Ni–CH). ^{13}C{^{1}H} NMR (75 MHz, pyridine-d_5): δ 181.5 (C=O), 137.6, 134.3, 129.4, 129.2, 129.0, 128.7 (PPh$_3$), 88.2, 80.4 (cage C), 49.8 (OCH_3), 42.4 (Ni–CH), 35.1 (CH$_2$). ^{11}B{^{1}H} NMR (96 MHz, pyridine-d_5): δ -3.1 (1B), -4.7 (1B), -8.1 (3B), -9.6 (4B), -13.6 (1B). IR (KBr, cm-1):ν 3055 (w), 2955 (w), 2880 (w), 2573 (vs), 1714 (m), 1619 (s), 1443 (s), 1274 (s), 1047 (s), 898 (m), 735 (m), 695 (m). Anal. Calcd for $C_{28}H_{39}B_{10}NiO_2P$ (**III-9r** + THF): C, 53.99; H, 6.32. Found: C, 53.75; H, 5.95.

Reaction of Complex III-9q with 3-Hexyne. Complex **III-9q** (20 mg, 0.018 mmol) was dissolved in THF (0.5 mL) and 3-hexyne (17 mg, 0.21 mmol) was then added. The reaction vessel was closed and heated at 110 °C for 3 days. After removal of the solvent, the residue was subject to column chromatography on silica gel (230–400 mesh) using hexane/ether (v/v = 12:1) as eluent to give **IV-1a** as a white solid (16 mg, 92%).

Reaction of Complex III-9r with 3-Hexyne. To a THF solution (0.5 mL) of **III-9r** (22 mg, 0.035 mmol) was added 3-hexyne (12 mg, 0.146 mmol). The reaction vessel was closed and heated at 110 °C for 3 days. Following the same workup procedure as described for **III-9q** gave **IV-1 h** as a white solid (10 mg, 91%).

8 Experimental Section

General Procedure for Nickel-Mediated Three-Component Reaction of 1,2-o-Carboryne with Alkenes and Alkynes. To a THF suspension (5 mL) of $Li_2C_2B_{10}H_{10}$ (1.0 mmol), prepared in situ from the reaction of ^nBuLi (2.0 mmol) with o-carborane (1.0 mmol), was added $NiCl_2(PPh_3)_2$ (1.0 mmol). The reaction mixture was stirred at room temperature for 0.5 h to give the Ni-1,2-o-carboryne complex [10]. Alkene (1.2 mmol) and alkyne (4.0 mmol) were added. The reaction vessel was closed, stirred at room temperature for 3 h and then heated at 110 °C for 3 days. After removal of the precipitate, the resulting solution was concentrated to dryness in vacuo. The residue was subject to column chromatography on silica gel to give the product **IV-1**.

1,2-[EtC=C(Et)CH(o-C$_5$H$_4$N)CH$_2$]-1,2-C$_2$B$_{10}$H$_{10}$ (IV-1a). Yield: 57%. White solid. ^1H NMR (400 MHz, CDCl$_3$): δ 8.59 (m, 1H), 7.70 (t, J = 7.4 Hz, 1H), 7.23 (m, 1H), 7.15 (d, J = 7.7 Hz, 1H) (Py), 3.93 (m, 1H) (CH), 2.89 (dd, J = 10.5, 14.6 Hz, 1H) (CHHCH), 2.75 (dd, J = 7.2, 14.6 Hz, 1H) (CHHCH), 2.43 (m, 1H), 2.34 (m, 1H), 2.25 (m, 1H), 1.53 (m, 1H) (CH_2CH$_3$), 1.15 (t, J = 7.4 Hz, 3H), 0.79 (t, J = 7.4 Hz, 3H) (CH$_2$CH_3). $^{13}C\{^1H\}$ NMR (100 MHz, CDCl$_3$): δ 161.1, 149.8, 137.0, 136.7, 130.8, 123.7, 122.1 (olefinic and Py), 73.4, 69.5 (cage C), 44.2 (CHCH_2), 37.0, (CHCH_2), 25.4, 24.1 (CH_2CH$_3$), 14.7, 12.6 (CH_3). $^{11}B\{^1H\}$ NMR (128 MHz, CDCl$_3$): δ −4.1 (1B), −5.5 (1B), −9.8 (2B), −10.7 (3B), −11.5 (3B). HRMS: m/z calcd for $C_{15}H_{27}N^{11}B_8^{10}B_2^+$: 329.3147. Found: 329.3149.

1,2-[^{n}BuC=C(^{n}Bu)CH(o-C$_5$H$_4$N)CH$_2$]-1,2-C$_2$B$_{10}$H$_{10}$ (IV-1b). Yield: 32%. Colorless oil. ^1H NMR (400 MHz, CDCl$_3$): δ 8.57 (d, J = 4.8 Hz,1H), 7.64 (t, J = 7.6 Hz, 1H), 7.18 (dd, J = 4.8, 7.6 Hz, 1H), 7.11 (d, J = 7.6 Hz, 1H) (Py), 3.84 (dd, J = 7.2, 10.8 Hz,1H) (CH), 2.84 (dd, J = 10.8, 14.4 Hz, 1H) (CHHCH), 2.69 (dd, J = 7.2, 14.4 Hz, 1H) (CHHCH), 2.38 (m, 1H), 2.21 (m, 3H), 1.63 (m, 1H), 1.48 (m, 1H), 1.36 (m, 3H), 1.24 (m, 1H), 1.10 (m, 2H) (CH_2), 0.94 (t, J = 7.2 Hz, 3H), 0.71 (t, J = 7.2 Hz, 3H) (CH_3). $^{13}C\{^1H\}$ NMR (100 MHz, CDCl$_3$): δ 161.2, 149.7, 136.7, 135.8, 130.0, 123.8, 122.1 (olefinic and Py), 73.6, 69.6 (cage C), 44.8 (CHCH_2), 37.2, (CHCH_2), 32.4, 32.2, 30.7, 30.1, 22.9, 22.4 (CH_2CH$_3$), 13.7, 13.6 (CH_3). $^{11}B\{^1H\}$ NMR (128 MHz, CDCl$_3$): δ −4.0 (1B), −5.4 (1B), −9.3 (2B), −10.7 (3B), −11.7 (3B). HRMS: m/z calcd for $C_{19}H_{35}N^{11}B_8^{10}B_2^+$: 385.3773. Found: 385.3765.

1,2-[^{i}PrC=C(Me)CH(o-C$_5$H$_4$N)CH$_2$]-1,2-C$_2$B$_{10}$H$_{10}$ (IV-1c) + 1,2-[MeC= C(^{i}Pr)CH(o-C$_5$H$_4$N)CH$_2$]-1,2-C$_2$B$_{10}$H$_{10}$ (IV-1c′). Yield: 34%. Colorless oil. IV-1c:IV-1c′ = 1.6:1. IV-1c: ^1H NMR (400 MHz, CDCl$_3$): δ 8.59 (m, 1H), 7.64 (m, 1H), 7.19 (m, 1H), 7.09 (d, J = 8.0 Hz, 1H) (aromatic H), 3.74 (m, 1H) (CHCH_2), 3.04 (m, 1H) (CH(CH$_3$)$_2$), 2.96 (dd, J = 10.4, 14.8 Hz, 1H), 2.73 (dd, J = 8.0, 14.8 Hz, 1H) (CHCH_2), 1.55 (s, 3H) (C=C–CH_3), 1.25 (d, J = 7.2 Hz, 6H) (CH(CH_3)$_2$). **IV-1c′**: ^1H NMR (400 MHz, CDCl$_3$): δ 8.56 (m, 1H), 7.64 (m, 1H), 7.17 (m, 2H) (aromatic H), 3.81 (m, 1H) (CHCH_2), 3.25–2.70 (m, 3H) (CH(CH$_3$)$_2$ & CHCH_2), 2.12 (s, 3H) (C=C–CH_3), 0.97 (d, J = 7.2 Hz, 3H), 0.80 (d, J = 7.2 Hz, 3H) (CH(CH_3)$_2$). Compound **IV-1c** and **IV-1c′** was isolated as a mixture and cannot be separated. Their molar ratio was determined by ^1H NMR spectrum on a crude product mixture.

1,2-[PhC=C(Me)CH(o-C₅H₄N)CH₂]-1,2-C₂B₁₀H₁₀ (**IV-1d**). Yield: 40%. White solid. ^{1}H NMR (400 MHz, CDCl₃): δ 8.60 (m, 1H), 7.68 (t, J = 7.6, 1H), 7.36 (m, 3H), 7.16 (m, 4H) (Py, Ph), 3.95 (dd, J = 7.2, 10.8 Hz, 1H) (C*H*CH₂), 3.11 (dd, J = 10.8, 14.8 Hz, 1H) (C*H*H), 2.84 (dd, J = 7.2, 14.8 Hz, 1H) (CH*H*), 1.15 (s, 3H) (=CC*H*₃). ^{13}C{^{1}H} NMR (100 MHz, CDCl₃): δ 160.1, 149.8, 137.8, 136.9, 134.5, 131.5, 130.1, 129.5, 128.4, 128.3, 127.9, 124.1, 122.4 (olefinic, Ph, and Py), 72.6, 70.0 (cage *C*), 46.3 (*C*HCH₂), 36.9 (CH*C*H₂), 20.4 (=C*C*H₃). ^{11}B{^{1}H} NMR (128 MHz, CDCl₃): δ −3.9 (1B), −5.4 (1B), −10.5 (8B). HRMS: m/z calcd for $C_{18}H_{25}N^{11}B_8^{10}B_2^+$: 363.2990. Found: 363.2995.

1,2-[(4′-Me-C₆H₄)C=C(Me)CH(o-C₅H₄N)CH₂]-1,2-C₂B₁₀H₁₀ (**IV-1e**). Yield: 35%. White solid. ^{1}H NMR (300 MHz, CDCl₃): δ 8.60 (m, 1H), 7.66 (t, J = 10.4 Hz, 1H), 7.17 (m, 4H), 7.03 (m, 2H) (Py, Ph), 3.93 (dd, J = 9.6, 14.8 Hz, 1H) (C*H*), 3.09 (dd, J = 14.8, 19.2 Hz, 1H) (C*H*H), 2.83 (dd, J = 9.6, 19.2 Hz, 1H) (CH*H*), 2.36 (s, 3H) (Ph–C*H*₃), 1.15 (s, 3H) (=CC*H*₃). ^{13}C{^{1}H} NMR (75 MHz, CDCl₃): δ 160.3, 149.9, 137.6, 136.8, 134.8, 134.5, 131.4, 129.9, 129.4, 129.1, 129.0, 124.0, 122.3 (olefinic, Ph, and Py), 72.8, 70.0 (cage *C*), 46.4 (*C*HCH₂), 36.9 (CH*C*H₂), 21.3 (Ph–*C*H₃), 20.4 (=C*C*H₃). ^{11}B{^{1}H} NMR (96 MHz, CDCl₃): δ −3.7 (1B), −5.1 (1B), −10.8 (8B). HRMS: m/z calcd for $C_{19}H_{27}N^{11}B_8^{10}B_2^+$: 377.3141. Found: 377.3143.

1,2-[PhC=C(Et)CH(o-C₅H₄N)CH₂]-1,2-C₂B₁₀H₁₀ (**IV-1f**). Yield: 39%. White solid. ^{1}H NMR (400 MHz, CDCl₃): δ 8.60 (m, 1H), 7.67 (t, J = 7.6 Hz, 1H), 7.37 (m, 3H), 7.16 (m, 4H) (Py, Ph), 4.03 (dd, J = 7.1, 10.8 Hz, 1H) (C*H*CH₂), 3.10 (dd, J = 10.8, 14.7 Hz, 1H) (C*H*HCH), 2.84 (dd, J = 7.1, 14.7 Hz, 1H) (CH*H*CH), 1.73 (m, 1H), 1.42 (m, 1H) (=CC*H*₂), 0.62 (t, J = 7.5 Hz, 3H) (C*H*₃). ^{13}C{^{1}H} NMR (75 MHz, CDCl₃): δ 160.2, 149.8, 140.2, 137.4, 136.7, 131.3, 130.0, 129.7, 128.2, 128.1, 127.9, 123.9, 122.3 (olefinic, Ph, and Py), 72.4, 69.7 (cage *C*), 44.0 (*C*HCH₂), 37.0 (CH*C*H₂), 25.9 (=C*C*H₂), 12.8 (*C*H₃). ^{11}B{^{1}H} NMR (96 MHz, CDCl₃): δ −4.3 (1B), −5.6 (1B), −10.9 (8B). HRMS: m/z calcd for $C_{19}H_{27}N^{11}B_8^{10}B_2^+$: 377.3141. Found: 377.3131.

1,2-[PhC=C(^{n}Bu)CH(o-C₅H₄N)CH₂]-1,2-C₂B₁₀H₁₀ (**IV-1g**). Yield: 31%. Colorless oil. ^{1}H NMR (400 MHz, CDCl₃): δ 8.60 (m, 1H), 7.67 (t, J = 7.6 Hz, 1H), 7.35 (m, 3H), 7.18 (m, 4H) (Py, Ph), 4.00 (dd, J = 7.1, 10.8 Hz, 1H) (C*H*CH₂), 3.08 (dd, J = 10.8, 14.6 Hz, 1H) (C*H*HCH), 2.83 (dd, J = 7.1, 14.6 Hz, 1H) (CH*H*CH), 1.66 (m, 1H), 1.39 (m, 1H) (=CC*H*₂), 1.08 (m, 1H), 0.96 (m, 1H), 0.83 (m, 1H), 0.79 (m, 1H) (C*H*₂), 0.50 (t, J = 7.3 Hz, 3H) (C*H*₃). ^{13}C{^{1}H} NMR (75 MHz, CDCl₃): δ 160.4, 149.8, 139.0, 137.5, 136.6, 131.6, 130.1, 129.9, 128.2, 128.0, 127.9, 124.0, 122.3 (olefinic, Ph, and Py), 72.5, 69.8 (cage *C*), 44.9 (*C*HCH₂), 37.1 (CH*C*H₂), 32.4, 30.2, 22.2 (*C*H₂), 13.4 (*C*H₃). ^{11}B{^{1}H} NMR (96 MHz, CDCl₃): δ −4.5 (1B), −5.9 (1B), −11.0 (8B). HRMS: m/z calcd for $C_{21}H_{31}N^{11}B_8^{10}B_2^+$: 405.3454. Found: 405.3442.

1,2-[PhC=C(CH₂CH=CH₂)CH(o-C₅H₄N)CH₂]-1,2-C₂B₁₀H₁₀ (**IV-1h**). Yield: 36%. Colorless oil. ^{1}H NMR (400 MHz, CDCl₃): δ 8.61 (m, 1H), 7.63 (m, 1H), 7.35 (m, 3H), 7.18 (m, 4H) (Py, Ph), 5.33 (m, 1H) (C*H*=CH₂), 4.87 (d, J = 10.2 Hz, 1H), 4.59 (d, J = 17.1 Hz, 1H) (CH=C*H*₂), 4.01 (dd, J = 7.1, 10.6 Hz, 1H) (cyclic C*H*CH₂), 3.16 (dd, J = 10.6, 14.7 Hz, 1H) (cyclic C*H*H),

8 Experimental Section

2.83 (dd, J = 7.1, 14.7 Hz, 1H) (cyclic CHH), 2.48 (dd, J = 4.8, 15.3 Hz, 1H), 2.15 (dd, J = 7.3, 15.3 Hz, 1H) (acyclic CH_2). ^{13}C{^{1}H} NMR (75 MHz, CDCl$_3$): δ 159.9, 149.9, 137.1, 136.5, 136.3, 134.5, 133.1, 129.8, 129.7, 128.1, 124.5, 122.3, 116.5 (olefinic, Ph, and Py), 72.3, 69.9 (cage C), 44.1 (CHCH$_2$), 36.8, 36.7 (CH$_2$). ^{11}B{^{1}H} NMR (96 MHz, CDCl$_3$): δ −3.4 (1B), −4.9 (1B), −10.2 (8B). HRMS: m/z calcd for C$_{20}$H$_{27}$N^{11}B$_8^{10}$B$_2^+$: 389.3141. Found: 389.3135.

1,2-[EtC=C(Et)CH(CO$_2$Me)CH$_2$]-1,2-C$_2$B$_{10}$H$_{10}$ (IV-1i). Yield: 59%. White solid. ^{1}H NMR (400 MHz, CDCl$_3$): δ 3.74 (s, 3H) (OCH$_3$), 3.28 (m, 1H) (CHCH$_2$), 3.13 (dd, J = 5.2, 14.8 Hz, 1H) (CHHCH), 2.54 (dd, J = 7.2, 14.8 Hz, 1H) (CHHCH), 2.46 (m, 2H), 2.29 (m, 1H), 2.05 (m, 1H), 1.11 (t, J = 7.6 Hz, 3H), 0.99 (t, J = 7.6 Hz, 3H) (CH$_3$). ^{13}C{^{1}H} NMR (100 MHz, CDCl$_3$): δ 172.4 (C=O), 133.3, 131.6 (olefinic), 73.2, 69.0 (cage C), 52.6 (OCH$_3$), 41.6 (CHCH$_2$), 32.6 (CHCH$_2$), 25.8, 25,2 (CH$_2$CH$_3$), 14.2, 12.7 (CH$_3$). ^{11}B{^{1}H} NMR (128 MHz, CDCl$_3$): δ −4.2 (1B), −5.7 (1B), −7.3 (1B), −9.2 (1B), −11.5 (6B). HRMS: m/z calcd for C$_{12}$H$_{26}$O$_2^{11}$B$_8^{10}$B$_2^+$: 310.2930. Found: 310.2922.

1,2-[^{n}PrC=C(^{n}Pr)CH(CO$_2$Me)CH$_2$]-1,2-C$_2$B$_{10}$H$_{10}$ (IV-1j). Yield: 50%. Colorless oil. ^{1}H NMR (400 MHz, CDCl$_3$): δ 3.74 (s, 3H) (OCH$_3$), 3.28 (m, 1H) (CHCH$_2$), 3.06 (dd, J = 6.0, 14.7 Hz, 1H) (CHHCH), 2.55 (dd, J = 7.3, 14.7 Hz, 1H) (CHHCH), 2.27 (m, 3H), 1.98 (m, 1H), 1.50 (m, 3H), 1.25 (m, 1H) (CH_2), 0.95 (t, J = 7.3 Hz, 3H), 0.91 (t, J = 7.3 Hz, 3H) (CH_3). ^{13}C{^{1}H} NMR (100 MHz, CDCl$_3$): δ 172.4 (C=O), 132.1, 130.8 (olefinic), 73.3, 69.0 (cage C), 52.6 (OCH$_3$), 42.2 (CHCH$_2$), 34.7, 34.4, 32.6, 23.0, 21.6 (CH$_2$), 14.1 (CH$_3$). ^{11}B{^{1}H} NMR (128 MHz, CDCl$_3$): δ −3.8 (1B), −5.3 (1B), −7.3 (1B), −8.3 (1B), −11.0 (6B). HRMS: m/z calcd for C$_{14}$H$_{30}$O$_2^{11}$B$_8^{10}$B$_2^+$: 338.3249. Found: 338.3237.

1,2-[^{n}BuC=C(^{n}Bu)CH(CO$_2$Me)CH$_2$]-1,2-C$_2$B$_{10}$H$_{10}$ (IV-1k). Yield: 48%. Colorless oil. ^{1}H NMR (400 MHz, CDCl$_3$): δ 3.73 (s, 3H) (OCH$_3$), 3.29 (m, 1H) (CHCH$_2$), 3.07 (dd, J = 6.1, 14.7 Hz, 1H) (CHHCH), 2.55 (dd, J = 7.3, 14.7 Hz, 1H) (CHHCH), 2.31 (m, 3H), 1.97 (m, 1H), 1.45 (m, 3H), 1.32 (m, 4H), 1.20 (m, 1H) (CH_2), 0.94 (t, J = 7.2 Hz, 3H), 0.91 (t, J = 7.2 Hz, 3H) (CH_3). ^{13}C{^{1}H} NMR (100 MHz, CDCl$_3$): δ 172.5 (C=O), 132.2, 130.6 (olefinic), 73.3, 69.0 (cage C), 52.6 (OCH$_3$), 42.2 (CHCH$_2$), 32.6, 32.5, 32.1, 31.7, 30.4, 22.9, 22.8 (CH$_2$), 13.9, 13.7 (CH$_3$). ^{11}B{^{1}H} NMR (128 MHz, CDCl$_3$): δ −3.8 (1B), −5.4 (1B), −7.1 (1B), −8.5 (1B), −11.0 (6B). HRMS: m/z calcd for C$_{16}$H$_{34}$O$_2^{11}$B$_8^{10}$B$_2^+$: 366.3562. Found: 366.3550.

General Procedure for Nickel-Catalyzed Three-Component Cyclization of Arynes with Alkenes and Alkynes. To a flask containing Ni(cod)$_2$ (0.015 mmol) and CsF (0.9 mmol) were added CH$_3$CN (1 mL), alkyne (0.6 mmol), alkene (0.6 mmol) and aryne precursor (0.3 mmol). The reaction mixture was stirred at room temperature for 5 h. After extraction with ether, the resulting solution was dried over Na$_2$SO$_4$ and concentrated to dryness in vacuo. The residue was subject to column chromatography on silica gel (40–230 mesh) using hexane/ethyl acetate as eluent to give the product.

1,2-[PhC=C(Ph)CH(CO$_2$Me)CH$_2$]C$_6$H$_4$ (IV-5a). Yield: 76%. Colorless oil. ^{1}H NMR (400 MHz, CDCl$_3$): δ 7.20 (m, 5H), 7.08 (m, 8H), 6.79 (d, J = 7.6 Hz,

1H) (aromatic *H*), 3.76 (t, *J* = 6.4 Hz, 1H) (C*H*), 3.56 (s, 3H) (OC*H*$_3$), 3.38 (dd, *J* = 6.4, 15.6 Hz, 1H) (CH*H*), 3.30 (dd, *J* = 6.4, 15.6 Hz, 1H) (CH*H*). ^{13}C{^{1}H} NMR (100 MHz, CDCl$_3$): δ 174.1 (*C*=O), 141.3, 139.1, 137.8, 136.1, 134.3, 133.3, 131.0, 129.8, 128.9, 127.9, 127.6, 127.3, 126.7, 126.3 (olefinic and aromatic *C*), 52.0 (O*C*H$_3$), 46.7 (*C*HCH$_2$), 32.3 (CH*C*H$_2$). HRMS: *m/z* calcd for C$_{24}$H$_{20}$O$_2$$^+$: 340.1458. Found: 340.1455.

1,2-[PhC=C(Ph)CH(CO$_2^n$Bu)CH$_2$]C$_6$H$_4$ (IV-5b). Yield: 72%. Colorless oil. ^{1}H NMR (300 MHz, CDCl$_3$): δ 7.20 (m, 5H), 7.08 (m, 8H), 6.78 (d, *J* = 6.9 Hz, 1H) (aromatic *H*), 3.97 (t, *J* = 6.3 Hz, 1H) (OC*H*$_2$), 3.72 (t, *J* = 6.0 Hz, 1H) (C*H*), 3.38 (dd, *J* = 6.0, 15.3 Hz, 1H) (C*H*H), 3.31 (dd, *J* = 6.0, 15.3 Hz, 1H) (CH*H*), 1.41 (m, 2H), 1.18 (m, 2H) (C*H*$_2$), 0.81 (t, *J* = 7.5 Hz, 3H) (C*H*$_3$). ^{13}C{^{1}H} NMR (75 MHz, CDCl$_3$): δ 173.7 (*C*=O), 141.4, 139.2, 137.6, 136.2, 134.4, 133.3, 131.0, 128.8, 127.9, 127.5, 127.3, 127.2, 126.6, 126.3 (olefinic and aromatic *C*), 64.6 (O*C*H$_2$), 46.9 (*C*HCH$_2$), 32.5 (CH*C*H$_2$), 30.5, 18.9 (*C*H$_2$), 13.6 (*C*H$_3$). HRMS: *m/z* calcd for C$_{27}$H$_{26}$O$_2$$^+$: 382.1927. Found: 382.1932.

1,2-[PhC=C(Ph)CH(CO$_2^t$Bu)CH$_2$]C$_6$H$_4$ (IV-5c). Yield: 74%. Colorless oil. ^{1}H NMR (400 MHz, CDCl$_3$): δ 7.17 (m, 5H), 7.08 (m, 8H), 6.75 (d, *J* = 7.2 Hz, 1H) (aromatic *H*), 3.64 (t, *J* = 6.4 Hz, 1H) (C*H*), 3.33 (dd, *J* = 6.4, 15.2 Hz, 1H) (C*H*H), 3.28 (dd, *J* = 6.4, 15.2 Hz, 1H) (CH*H*), 1.21 (s, 9H) (C(C*H*$_3$)$_3$). ^{13}C{^{1}H} NMR (75 MHz, CDCl$_3$): δ 172.9 (*C*=O), 141.4, 139.3, 137.2, 136.5, 135.1, 133.6, 131.0, 128.8, 127.9, 127.5, 127.2, 127.1, 126.6, 126.5, 126.4, 126.2 (olefinic and aromatic *C*), 80.7 (O*C*(CH$_3$)$_3$), 48.1 (*C*HCH$_2$), 32.8 (CH*C*H$_2$), 27.6 (*C*H$_3$). HRMS: *m/z* calcd for C$_{27}$H$_{26}$O$_2$$^+$: 382.1927. Found: 382.1921.

1,2-{PhC=C(Ph)[CHC(=O)Me]CH$_2$}C$_6$H$_4$ (IV-5d). Yield: 3%. Colorless oil. ^{1}H NMR (300 MHz, CDCl$_3$): δ 7.24 (m, 4H), 7.13 (m, 7H), 7.05 (m, 2H), 6.83 (d, *J* = 7.5 Hz, 1H) (aromatic *H*), 3.67 (dd, *J* = 4.8, 6.9 Hz, 1H) (C*H*), 3.41 (dd, *J* = 6.9, 15.6 Hz, 1H) (C*H*H), 3.20 (dd, *J* = 4.8, 15.6 Hz, 1H) (CH*H*), 2.10 (s, 1H), (C*H*$_3$). ^{13}C{^{1}H} NMR (75 MHz, CDCl$_3$): δ 209.9 (*C*=O), 141.4, 139.0, 138.1, 136.2, 134.8, 133.0, 130.9, 128.8, 128.0, 127.8, 127.5, 127.3, 126.8, 126.7, 126.5 (olefinic and aromatic *C*), 54.8 (*C*HCH$_2$), 32.4 (CH*C*H$_2$), 28.7 6 (*C*H$_3$). HRMS: *m/z* calcd for C$_{24}$H$_{20}$O$_2$$^+$: 324.1509. Found: 324.1498.

1,2-[PhC=C(Ph)CH(CN)CH$_2$]C$_6$H$_4$ (IV-5e). Yield: 15%. Colorless crystals. ^{1}H NMR (400 MHz, CDCl$_3$): δ 7.24 (m, 5H), 7.16 (m, 6H), 7.12 (m, 2H), 6.84 (d, *J* = 8.0 Hz, 1H) (aromatic *H*), 3.84 (dd, *J* = 4.4, 5.6 Hz, 1H) (C*H*), 3.40 (dd, *J* = 5.6, 15.2 Hz, 1H) (C*H*H), 3.20 (dd, *J* = 4.4, 15.2 Hz, 1H) (CH*H*). ^{13}C{^{1}H} NMR (100 MHz, CDCl$_3$): δ 139.8, 139.2, 138.1, 135.3, 131.3, 130.7, 129.5, 128.2, 128.1, 127.7, 127.5, 127.3, 127.2 (olefinic and aromatic *C*), 120.4 (*C*N), 32.4, 32.2 (*C*HCH$_2$ & CH*C*H$_2$). HRMS: *m/z* calcd for C$_{23}$H$_{17}$N$^+$: 307.1356. Found: 307.1361.

1,2-(OCH$_2$O)-4,5-[PhC=C(Ph)CH(CO$_2$Me)CH$_2$]C$_6$H$_2$ (IV-5f). Yield: 29%. Colorless oil. ^{1}H NMR (400 MHz, CDCl$_3$): δ 7.20 (m, 3H), 7.07 (m, 7H), 6.74 (s, 1H), 6.31 (s, 1H) (aromatic *H*), 5.88 (s, 2H) (OC*H*$_2$O), 3.67 (t, *J* = 6.0 Hz, 1H) (C*H*), 3.60 (s, 3H) (OC*H*$_3$), 3.28 (dd, *J* = 6.0, 15.6 Hz, 1H) (C*H*H), 3.23 (dd, *J* = 6.0, 15.6 Hz, 1H) (CH*H*). ^{13}C{^{1}H} NMR (75 MHz, CDCl$_3$): δ 174.2 (*C*=O), 146.3, 146.1, 141.4, 139.3, 137.5, 132.3, 130.9, 130.4, 128.8, 127.9, 127.5,

8 Experimental Section

127.4, 126.7, 126.1, 108.1, 107.7 (olefinic and aromatic C), 100.9 (OCH_2O), 52.1 (OCH_3), 46.7 ($CHCH_2$), 32.4 ($CHCH_2$). HRMS: m/z calcd for $C_{25}H_{20}O_4^+$: 384.1356. Found: 384.1350.

4,5-(CH_2)$_3$-1,2-[PhC=C(Ph)CH(CO$_2$Me)CH$_2$]C$_6$H$_4$ (IV-5 g). Yield: 46%. Colorless oil. ^{1}H NMR (400 MHz, CDCl$_3$): δ 7.20 (m, 3H), 7.07 (m, 8H), 6.64 (s, 1H) (aromatic H), 3.73 (t, $J = 6.0$ Hz, 1H) (CH), 3.57 (s, 3H) (OCH_3), 3.35 (dd, $J = 6.0$, 15.6 Hz, 1H) (CHH), 3.26 (dd, $J = 6.0$, 15.6 Hz, 1H) (CHH), 2.88 (t, $J = 7.6$ Hz, 2H), 2.73 (t, $J = 7.6$ Hz, 2H) ($CH_2CH_2CH_2$), 2.01 (m, 2H) ($CH_2CH_2CH_2$). ^{13}C$\{^1$H$\}$ NMR (100 MHz, CDCl$_3$): δ 174.3 (C=O), 143.5, 142.5, 141.5, 139.5, 138.2, 134.3, 133.2, 131.3, 131.0, 128.9, 127.8, 127.5, 126.5, 126.1, 123.4, 122.7 (olefinic and aromatic C), 52.0 (OCH_3), 47.0 ($CHCH_2$), 32.7, 32.6 ($CHCH_2$ & $CH_2CH_2CH_2$), 25.4 ($CH_2CH_2CH_2$). HRMS: m/z calcd for $C_{27}H_{24}O_2^+$: 380.1771. Found: 380.1778.

4-Me-1,2-[PhC=C(Ph)CH(CO$_2$Me)CH$_2$]C$_6$H$_4$ (IV-5 h) + 5-Me-1,2-[PhC =C(Ph)CH(CO$_2$Me)CH$_2$]C$_6$H$_4$ (IV-5′h): Yield: 57%. Colorless oil. ^{1}H NMR (400 MHz, CDCl$_3$): δ 7.22 (m, 6H), 7.04 (m, 16H), 6.97 (d, $J = 8.0$ Hz, 1H), 6.88 (d, $J = 8.0$ Hz, 1H), 6.67 (d, $J = 8.0$ Hz, 1H), 6.60 (s, 1H) (aromatic H), 3.73 (m, 2H) (CH), 3.57 (s, 3H), 3.56 (s, 3H) (OCH_3), 3.34 (m, 2H) (CHH), 3.26 (m, 2H) (CHH), 2.32 (s, 3H), 2.18 (s, 3H) (CH_3). ^{13}C$\{^1$H$\}$ NMR (100 MHz, CDCl$_3$): δ 174.2 (C=O), 141.4, 139.3, 139.2, 137.9, 137.7, 137.2, 136.1, 135.9, 134.3, 133.2, 131.0, 130.2, 129.6, 128.9, 128.8, 128.2, 128.0, 127.8, 127.5, 127.4, 127.3, 127.2, 126.6, 126.2, 126.1 (olefinic and aromatic C), 52.0 (OCH_3), 46.9, 46.7 ($CHCH_2$), 32.3, 31.9 ($CHCH_2$), 21.3, 21.2 (CH_3). HRMS: m/z calcd for $C_{25}H_{22}O_2^+$: 354.1614. Found: 354.1606.

1,2-[MeC=C(Ph)CH(CO$_2$Me)CH$_2$]C$_6$H$_4$ (IV-5i). Yield: 71%. Colorless oil. ^{1}H NMR (400 MHz, CDCl$_3$): δ 7.34 (m, 3H), 7.27 (m, 4H), 7.17 (m, 2H) (aromatic H), 3.57 (m, 1H) (CH), 3.43 (s, 3H) (OCH_3), 3.29 (dd, $J = 4.7$, 15.5 Hz, 1H) (CHH), 3.19 (dd, $J = 7.0$, 15.5 Hz, 1H) (CHH), 2.00 (s, 3H) (CH_3). ^{13}C$\{^1$H$\}$ NMR (100 MHz, CDCl$_3$): δ 173.5 (C=O), 142.0, 135.8, 133.6, 133.3, 130.3, 129.1, 128.1, 127.4, 127.1, 126.8, 123.9 (olefinic and aromatic C), 51.7 (OCH_3), 46.1 ($CHCH_2$), 32.0 ($CHCH_2$), 16.2 (CH_3). HRMS: m/z calcd for $C_{19}H_{18}O_2^+$: 278.1301. Found: 278.1299.

1,2-[EtC=C(Ph)CH(CO$_2$Me)CH$_2$]C$_6$H$_4$ (IV-5j). Yield: 78%. Colorless oil. ^{1}H NMR (400 MHz, CDCl$_3$): δ 7.26 (m, 3H), 7.23 (m, 4H), 7.17 (m, 2H) (aromatic H), 3.52 (t, $J = 6.4$ Hz, 1H) (CH), 3.43 (s, 3H) (OCH_3), 3.25 (dd, $J = 6.4$, 15.4 Hz, 1H) ($CHCHH$), 3.12 (dd, $J = 6.4$, 15.4 Hz, 1H) ($CHCHH$), 2.46 (q, $J = 7.5$ Hz, 2H) (CH_2CH_3), 0.99 (t, $J = 7.5$ Hz, 3H) (CH_2CH_3). ^{13}C$\{^1$H$\}$ NMR (75 MHz, CDCl$_3$): δ 173.8 (C=O), 141.9, 136.4, 134.1, 133.9, 133.3, 128.6, 128.1, 127.7, 126.9, 126.8, 123.9 (olefinic and aromatic C), 51.7 (OCH_3), 46.4 ($CHCH_2$), 32.2 ($CHCH_2$), 22.0 (CH_2CH_3), 14.1 (CH_2CH_3). HRMS: m/z calcd for $C_{20}H_{20}O_2^+$: 292.1458. Found: 292.1461.

1,2-[^{n}BuC=C(Ph)CH(CO$_2$Me)CH$_2$]C$_6$H$_4$ (IV-5 k). Yield: 71%. Colorless oil. ^{1}H NMR (400 MHz, CDCl$_3$): δ 7.28 (m, 3H), 7.22 (m, 4H), 7.18 (m, 2H) (aromatic H), 3.54 (m, 1H) (CH), 3.44 (s, 3H) (OCH_3), 3.27 (dd, $J = 5.7$, 15.4 Hz, 1H) ($CHCHH$), 3.14 (dd, $J = 6.8$, 15.4 Hz, 1H) ($CHCHH$), 2.45 (t, $J = 7.8$ Hz,

2H) (CH_2 CH$_2$), 1.36 (m, 2H), 1.21 (m, 2H) (CH_2), 0.76 (t, $J = 7.2$ Hz, 3H) (CH$_2$CH_3). ^{13}C{^{1}H} NMR (100 MHz, CDCl$_3$): δ 173.7 (C=O), 142.0, 135.0, 134.2, 134.1, 133.7, 128.8, 128.1, 127.7, 126.9, 126.7, 124.0 (olefinic and aromatic C), 51.7 (OCH_3), 46.4 (CHCH_2), 32.2 (CHCH_2), 31.2, 28.5, 22.5 (CH_2), 13.8 (CH$_2$CH_3). HRMS: m/z calcd for C$_{22}$H$_{24}$O$_2^+$: 320.1771. Found: 320.1777.

1,2-[C(CH$_2$OMe)=C(Ph)CH(CO$_2$Me)CH$_2$]C$_6$H$_4$ (IV-5 l). Yield: 75%. Colorless crystals. ^{1}H NMR (400 MHz, CDCl$_3$): δ 7.43 (d, $J = 7.6$ Hz, 1H), 7.24 (m, 5H), 7.17 (m, 1H), 7.10 (m, 2H) (aromatic H), 4.17 (d, $J = 10.8$ Hz, 1H) (CHHOCH$_3$), 4.10 (d, $J = 10.8$ Hz, 1H) (CHHOCH$_3$), 3.55 (dd, $J = 4.4, 6.8$ Hz, 1H) (CH), 3.37 (s, 3H) (CO$_2$CH_3), 3.24 (dd, $J = 4.4, 15.5$ Hz, 1H) (CHCHH), 3.19 (s, 3H) (CH$_2$OCH_3), 3.13 (dd, $J = 6.8, 15.5$ Hz, 1H) (CHCHH). ^{13}C{^{1}H} NMR (100 MHz, CDCl$_3$): δ 173.0 (C=O), 140.7, 138.2, 133.6, 133.3, 131.3, 128.9, 128.0, 127.5, 127.4, 127.3, 127.0, 124.7 (olefinic and aromatic C), 69.3 (OCH_2), 57.7 (CH$_2$OCH_3), 51.9 (CO$_2$CH_3), 46.2 (CHCH_2), 31.8 (CHCH_2). HRMS: m/z calcd for C$_{20}$H$_{20}$O$_3^+$: 308.1407. Found: 308.1415.

1,2-[C(CH$_2$CH=CH$_2$)=C(Ph)CH(CO$_2$Me)CH$_2$]C$_6$H$_4$ (IV-5 m). Yield: 68%. Colorless oil. ^{1}H NMR (400 MHz, CDCl$_3$): δ 7.30 (m, 6H), 7.19 (m, 3H) (aromatic H), 5.90 (m, 1H) (CH=CH$_2$), 5.05 (m, 2H) (CH=CH_2), 3.59 (dd, $J = 5.4, 6.6$ Hz, 1H) (CH), 3.48 (s, 3H) (OCH_3), 3.30 (dd, $J = 5.2, 15.5$ Hz, 1H) (CHCHH), 3.21 (m, 3H) (CHCHH & CH_2CH=CH$_2$). ^{13}C{^{1}H} NMR (100 MHz, CDCl$_3$): δ 173.6 (C=O), 141.7, 136.8, 135.2, 134.3, 133.6, 132.1, 128.4, 128.1, 127.5, 127.1, 127.0, 126.7, 124.7, 116.0 (olefinic and aromatic C), 51.8 (OCH_3), 46.4 (CHCH_2), 33.5, 31.8 (CHCH_2 & CH_2CH=CH$_2$). HRMS: m/z calcd for C$_{21}$H$_{20}$O$_2^+$: 304.1458. Found: 304.1457.

1,2-{C[(CH$_2$)$_3$CN]=C(Ph)CH(CO$_2$Me)CH$_2$}C$_6$H$_4$ (IV-5n). Yield: 32%. White solid. ^{1}H NMR (400 MHz, CDCl$_3$): δ 7.23 (m, 9H) (aromatic H), 3.52 (dd, $J = 5.3, 6.8$ Hz, 1H) (CH), 3.43 (s, 3H) (OCH_3), 3.26 (dd, $J = 5.3, 15.5$ Hz, 1H) (CHCHH), 3.14 (dd, $J = 6.8, 15.5$ Hz, 1H) (CHCHH), 2.60 (m, 2H), 2.13 (m, 2H), 1.72 (m, 2H) (CH_2). ^{13}C{^{1}H} NMR (100 MHz, CDCl$_3$): δ 173.2 (C=O), 141.3, 135.8, 134.0, 133.3, 132.7, 128.7, 128.4, 128.0, 127.4, 127.2, 127.0, 123.6 (olefinic and aromatic C), 119.4 (CN), 51.8 (OCH_3), 46.4 (CHCH_2), 32.0 (CHCH_2), 27.6, 24.6, 16.7 (CH_2). HRMS: m/z calcd for C$_{22}$H$_{21}$NO$_2^+$: 331.1567. Found: 331.1564.

1,2-[MeC=C(4$'$-Me-C$_6$H$_4$)CH(CO$_2$Me)CH$_2$]C$_6$H$_4$ (IV-5o). Yield: 66%. Colorless oil. ^{1}H NMR (400 MHz, CDCl$_3$): δ 7.23 (m, 1H), 7.18 (m, 1H), 7.17 (m, 6H) (aromatic H), 3.56 (dd, $J = 4.4, 6.8$ Hz, 1H) (CH), 3.44 (s, 3H) (OCH_3), 3.29 (dd, $J = 4.4, 15.2$ Hz, 1H) (CHH), 3.18 (dd, $J = 6.8, 15.2$ Hz, 1H) (CHH), 2.36 (s, 3H) (C$_6$H$_4$–CH_3), 2.02 (s, 3H) (C=C–CH_3). ^{13}C{^{1}H} NMR (75 MHz, CDCl$_3$): δ 173.6 (C=O), 139.0, 136.4, 135.9, 133.4, 133.3, 130.1, 129.0, 128.7, 127.4, 126.9, 126.8, 123.8 (olefinic and aromatic C), 51.7 (OCH_3), 46.0 (CHCH_2), 32.0 (CHCH_2), 21.2 (C$_6$H$_4$–CH_3), 16.3 (C=C–CH_3). HRMS: m/z calcd for C$_{20}$H$_{20}$O$_2^+$: 292.1458. Found: 292.1457.

1,2-[MeC=C(CO$_2$Me)CH(CO$_2$Me)CH$_2$]C$_6$H$_4$ (IV-5p). Yield: 19%. Colorless oil. ^{1}H NMR (400 MHz, CDCl$_3$): δ 7.48 (m, 1H), 7.26 (m, 2H), 7.18 (m, 1H) (aromatic H), 3.91 (dd, $J = 4.0, 7.2$ Hz, 1H) (CH), 3.81 (s, 3H), 3.56 (s, 3H) (OCH_3), 3.24 (dd, $J = 4.0, 15.6$ Hz, 1H) (CHH), 3.05 (dd, $J = 7.2, 15.6$ Hz, 1H)

8 Experimental Section

(CH*H*), 2.51 (s, 3H) (s, 3H) (C=C–CH$_3$). ^{13}C{^{1}H} NMR (75 MHz, CDCl$_3$): δ 173.3, 168.4 (*C*=O), 143.7, 135.0, 134.5, 129.1, 127.6, 127.0, 126.6, 125.4 (olefinic and aromatic *C*), 52.1, 51.7 (O*C*H$_3$), 40.5 (CH*C*H$_2$), 30.9 (CH*C*H$_2$), 16.8 (*C*H$_3$). HRMS: *m/z* calcd for C$_{15}$H$_{16}$O$_4^+$: 260.1043. Found: 260.1040.

1,2-[EtC=C(Et)CH(CO$_2$Me)CH$_2$]C$_6$H$_4$ (IV-5q). 3 mmol of 3-hexyne was used in the reaction. Yield: 63%. Colorless oil. ^{1}H NMR (300 MHz, CDCl$_3$): δ 7.15 (m, 4H) (aromatic *H*), 3.56 (s, 3H) (O*C*H$_3$), 3.21 (m, 2H) (C*H* & C*H*H), 2.96 (dd, *J* = 6.9, 15.3 Hz, 1H) (CH*H*), 2.56 (m, 3H), 2.16 (m, 1H) (C*H$_2$*CH$_3$), 1.11 (t, *J* = 7.5 Hz, 3H), 1.10 (t, *J* = 7.5 Hz, 3H) (CH$_2$*C*H$_3$). ^{13}C{^{1}H} NMR (75 MHz, CDCl$_3$): δ 174.0 (*C*=O), 134.7, 134.1, 133.8, 133.7, 127.5, 126.6, 126.1, 123.0 (olefinic and aromatic *C*), 51.8 (O*C*H$_3$), 42.7 (CH*C*H$_2$), 32.0 (CH*C*H$_2$), 26.1, 20.9 (*C*H$_2$CH$_3$), 14.1, 13.6 (CH$_2$*C*H$_3$). HRMS: *m/z* calcd for C$_{16}$H$_{20}$O$_2^+$: 244.1458. Found: 244.1455.

1,2-[^{n}BuC=C(^{n}Bu)CH(CO$_2$Me)CH$_2$]C$_6$H$_4$ (IV-5r). 3 mmol of 5-decyne was used in the reaction. Yield: 47%. Colorless oil. ^{1}H NMR (300 MHz, CDCl$_3$): δ 7.19 (m, 4H) (aromatic *H*), 3.56 (s, 3H) (O*C*H$_3$), 3.12 (m, 2H) (C*H* & C*H*H), 2.93 (dd, *J* = 6.9, 15.3 Hz, 1H) (CH*H*), 2.54 (m, 3H), 2.09 (m, 1H), 1.43 (m, 8H) (*C*H$_2$CH$_3$), 0.93 (m, 6H) (CH$_2$*C*H$_3$). ^{13}C{^{1}H} NMR (75 MHz, CDCl$_3$): δ 174.0 (*C*=O), 135.0, 133.8, 133.1, 132.7, 127.5, 126.5, 126.1, 123.1 (olefinic and aromatic *C*), 51.8 (O*C*H$_3$), 43.1 (CH*C*H$_2$), 32.9 (CH*C*H$_2$), 31.8, 31.6, 31.3, 27.8, 23.1, 23.0 (*C*H$_2$), 14.1 (CH$_2$*C*H$_3$). HRMS: *m/z* calcd for C$_{20}$H$_{28}$O$_2^+$: 300.2084. Found: 300.2079.

1,2-[^{i}PrC=C(Me)CH(CO$_2$Me)CH$_2$]C$_6$H$_4$ (IV-5s) + 1,2-[MeC=C(^{i}Pr)CH (CO$_2$Me)CH$_2$]C$_6$H$_4$ (IV-5 s$'$). Yield: 51%. Colorless oil, **IV-5s:IV-5s$'$** = 2:1. **IV-5s:** ^{1}H NMR (400 MHz, CDCl$_3$): δ 7.41–7.07 (m, 4H) (aromatic *H*), 3.50 (s, 3H) (O*C*H$_3$), 3.23–2.89 (m, 4H) (CH*C*H$_2$, CHC*H$_2$*, & C*H*(CH$_3$)$_2$), 2.13 (s, 3H) (C=C–C*H$_3$), 1.10 (d, *J* = 7.2 Hz, 3H), 0.99 (d, *J* = 7.2 Hz, 3H) (CH(C*H$_3$*)$_2$). **IV-5s$'$:** ^{1}H NMR (400 MHz, CDCl$_3$): δ 7.41–7.07 (m, 4H) (aromatic *H*), 3.59 (s, 3H) (O*C*H$_3$), 3.23–2.89 (m, 4H) (CH*C*H$_2$, CHC*H$_2$*, & C*H*(CH$_3$)$_2$), 2.02 (s, 3H) (C=C–C*H$_3$), 1.35 (d, *J* = 7.2 Hz, 3H), 1.30 (d, *J* = 7.2 Hz, 3H) (CH(C*H$_3$*)$_2$). Compounds **IV-5s** and **IV-5s$'$** were isolated as a mixture and cannot be separated. Their molar ratio was determined by ^{1}H NMR spectrum on a crude product mixture.

Control Experiment: Reaction of 1,2-*o*-carboryne precursor with toluene. To a toluene solution (10 mL) of Li$_2$C$_2$B$_{10}$H$_{10}$, prepared in situ from the reaction of ^{n}BuLi (1.6 M, 1.25 mL, 2.0 mmol) with *o*-carborane (144 mg, 1.0 mmol), was added I$_2$ (254 mg, 1.0 mmol). The reaction mixture was stirred at room temperature for 0.5 h and then heated at 110 °C for 2 h. After addition of water and extraction with ether, the resulting solution was concentrated to dryness in vacuo. The residue was subject to column chromatography on silica gel (230–400 mesh) using hexane as eluent to give **V-2** as a white solid (89 mg, 38%).

1,2-(2-methyl-2,5-cyclohexadiene-1,4-diyl)-*o*-carborane (V-2a).: ^{1}H NMR (400 MHz, CDCl$_3$): δ 6.67 (m, 2H), 6.22 (dd, *J* = 1.6, 4.4 Hz, 1H) (olefinic *H*), 4.03 (td, *J* = 1.6, 6.4 Hz, 1H), 3.78 (d, *J* = 5.6 Hz, 1H) (C*H*), 1.87 (s, 3H) (C*H$_3$*).

116   8 Experimental Section

1,2-(1-methyl-2,5-cyclohexadiene-1,4-diyl)-o-carborane (**V-2b**): ^{1}H NMR (400 MHz, CDCl$_3$): δ 6.67 (m, 2H), 6.36 (d, $J = 7.6$ Hz, 2H) (olefinic H), 4.07 (m, 1H) (CH), 1.64 (s, 3H) (CH_3).

Control Experiment: Reaction of 1,2-o-carboryne precursor with 3-hexyne in toluene. To a toluene solution (10 mL) of Li$_2$C$_2$B$_{10}$H$_{10}$, prepared in situ from the reaction of ^{n}BuLi (1.6 M, 1.25 mL, 2.0 mmol) with o-carborane (144 mg, 1.0 mmol), was added I$_2$ (254 mg, 1.0 mmol). The reaction mixture was stirred at room temperature for 0.5 h. 3-Hexyne (328 mg, 4.0 mmol) was then added and the reaction vessel was closed and then heated at 110 °C for 2 h. After addition of water and extraction with ether, the resulting solution was concentrated to dryness in vacuo. The residue was subject to column chromatography on silica gel (230–400 mesh) using hexane as eluent to give [4 + 2] cycloaddition product **V-2** as a white solid (40 mg, 17%), the ene-reaction product **V-3** as a colorless oil (81 mg, 36%), and carborane as a white solid (39 mg, 27%). **1-[C(Et)= C=CH(Me)]-1,2-C$_2$B$_{10}$H$_{11}$** (**V-3**). ^{1}H NMR (400 MHz, CDCl$_3$): δ 5.53 (m, 1H) (CH), 3.75 (s, 1H) (cage H), 2.08 (m, 2H) (CH_2), 1.72 (d, $J = 7.2$ Hz, 3H) (CHCH_3), 0.96 (t, $J = 7.2$ Hz, 3H) (CH$_2$CH_3). ^{13}C{^{1}H} NMR (75 MHz, CDCl$_3$): δ 202.6 (C=C=CH), 104.8, 94.0 (olefinic C), 74.6, 60.7 (cage C), 24.7 (CH), 14.0, 12.2 (CH$_3$). ^{11}B{^{1}H} NMR (96 MHz, CDCl$_3$): δ -3.6 (1B), -6.1 (1B), -10.4 (2B), -11.6 (2B), -13.3 (2B), -14.2 (2B). HRMS: m/z Calcd for [M-2H]$^+$ (C$_8$H$_{20}$B$_{10}$$^+$): 222.2406. Found: 222.2403.

Control Experiment: Heating of 1,2-o-Carboryne precursor in toluene in the presence of NiCl$_2$(PPh$_3$)$_2$. To a toluene solution (10 mL) of Li$_2$C$_2$B$_{10}$H$_{10}$, prepared in situ from the reaction of ^{n}BuLi (1.6 M, 1.25 mL, 2.0 mmol) with o-carborane (144 mg, 1.0 mmol), was added I$_2$ (254 mg, 1.0 mmol). The reaction mixture was stirred at room temperature for 0.5 h. NiCl$_2$(PPh$_3$)$_2$ (654 mg, 1.0 mmol) was then added and the reaction vessel was closed and then heated at 110 °C for 2 h. After addition of water and extraction with ether, the resulting solution was concentrated to dryness in vacuo. The residue was subject to column chromatography on silica gel (230–400 mesh) using hexane as eluent to give 1-iodocarborane **V-4** as a white solid (41 mg, 15%), and carborane as a white solid (76 mg, 53%).

General Procedure for Nickel-Catalyzed Regioselective [2+2+2] Cycloaddition of 1,2-o-Carboryne with Alkynes or Diynes. To a toluene solution (5 mL) of Li$_2$C$_2$B$_{10}$H$_{10}$ (0.5 mmol), prepared in situ from the reaction of ^{n}BuLi (1.0 mmol) with o-carborane (0.5 mmol), was added I$_2$ (0.5 mmol), and the reaction mixture was stirred at room temperature for 0.5 h. NiCl$_2$(PPh$_3$)$_2$ (0.1 mmol), and alkyne (2.0 mmol) or diyne (1.0 mmol) were then added, and the reaction vessel was closed and heated at 110 °C overnight. After addition of 5 mL of water and extraction with ether (10 mL $\times$ 3), the resulting ether solutions were concentrated to dryness in vacuo. The residue was subject to flash column chromatography on silica gel (230–400 mesh) using hexane as eluent to give the cycloaddition product.

1,2-[EtC=C(Et)C(Et)=CEt]-1,2-C$_2$B$_{10}$H$_{10}$ (**V-1a**). Yield: 65%. Colorless crystals. ^{1}H NMR (300 MHz, CDCl$_3$): δ 2.42 (q, $J = 7.5$ Hz, 4H), 2.01 (q, $J = 7.5$ Hz, 4H) (CH_2), 1.02 (t, $J = 7.5$ Hz, 6H), 0.78 (t, $J = 7.5$ Hz, 6H)

8 Experimental Section

(CH_3). $^{13}C\{^1H\}$ NMR (75 MHz, CDCl$_3$): δ 135.1, 134.0 (olefinic C), 76.3 (cage C), 26.3, 21.9 (CH_2CH_3), 15.0, 14.8 (CH_2CH_3). $^{11}B\{^1H\}$ NMR (96 MHz, CDCl$_3$): δ −7.4 (2B), −10.2 (6B), −13.1 (2B). These data are identical with those reported in the literature [10].

3-Cl-1,2-[EtC=C(Et)C(Et)=CEt]-1,2-C$_2$B$_{10}$H$_9$ (V-1b). Yield: 31%. White solid. ^1H NMR (400 MHz, CDCl$_3$): δ 2.51 (q, $J = 7.4$ Hz, 4H), 2.37 (q, $J = 7.4$ Hz, 4H) (CH_2), 1.15 (t, $J = 7.4$ Hz, 6H), 1.05 (t, $J = 7.4$ Hz, 6H) (CH_3). $^{13}C\{^1H\}$ NMR (75 MHz, CDCl$_3$): δ 136.9, 132.6 (olefinic C), 26.4, 22.2 (CH_2), 15.1, 14.6 (CH_3), cage C atoms were not observed. $^{11}B\{^1H\}$ NMR (96 MHz, CDCl$_3$): δ −8.9 (4B), −10.7 (2B), −12.5 (2B), −14.0 (1B), −17.9 (1B). HRMS: m/z Calcd for $C_{14}H_{29}B_{10}Cl^+$: 340.2955. Found: 340.2954.

3-Ph-1,2-[EtC=C(Et)C(Et)=CEt]-1,2-C$_2$B$_{10}$H$_9$ (V-1c). Yield: 38%. White solid. ^1H NMR (400 MHz, CDCl$_3$): δ 7.30 (m, 2H), 7.23 (m, 1H), 7.18 (m, 2H) (Ph), 2.63 (m, 4H), 2.06 (m, 4H) (CH_2), 1.19 (t, $J = 7.2$ Hz, 6H), 0.56 (t, $J = 7.2$ Hz, 6H) (CH_3). $^{13}C\{^1H\}$ NMR (100 MHz, CDCl$_3$): δ 135.5, 134.1, 133.7, 128.9, 127.3 (Ph & olefinic C), 26.5, 21.8 (CH_2), 15.0, 14.0 (CH_3), cage C atoms were not observed. $^{11}B\{^1H\}$ NMR (96 MHz, CDCl$_3$): δ −5.9 (1B), −7.7 (2B), −11.1 (3B), −12.0 (3B), −15.4 (1B). HRMS: m/z Calcd for $C_{20}H_{34}B_{10}^+$: 382.3658. Found: 382.3658.

1,2-[^{n}PrC=C(^{n}Pr)C(^{n}Pr)=C^{n}Pr]-1,2-C$_2$B$_{10}$H$_{10}$ (V-1d). Yield: 59%. White solid. ^1H NMR (400 MHz, CDCl$_3$): δ 2.45 (m, 4H), 2.17 (m, 4H), 1.53 (m, 4H), 1.32 (m, 4H) (CH_2), 0.97 (m, 12H, CH_3). $^{13}C\{^1H\}$ NMR (100 MHz, CDCl$_3$): δ 134.1, 132.8 (olefinic C), 76.4 (cage C), 35.8, 31.4, 23.9, 23.8 (CH_2), 14.3 (CH_3). $^{11}B\{^1H\}$ NMR (96 MHz, CDCl$_3$): δ −8.3 (2B), −11.2 (6B), −14.0 (2B). These data are identical with those reported in the literature [10].

1,2-[^{n}BuC=C(^{n}Bu)C(^{n}Bu)=C^{n}Bu]-1,2-C$_2$B$_{10}$H$_{10}$ (V-1e). Yield: 54%. White solid. ^1H NMR (400 MHz, CDCl$_3$): δ 2.49 (m, 4H), 2.20 (m, 4H), 1.45 (m, 4H), 1.39 (m, 8H), 1.24 (m, 4H) (CH_2), 0.95 (m, 12H, CH_3). $^{13}C\{^1H\}$ NMR (100 MHz, CDCl$_3$): δ 134.0, 132.8 (olefinic C), 76.5 (cage C), 33.4, 32.6, 29.0, 23.1, 23.0 (CH_2), 13.8, 13.7 (CH_3). $^{11}B\{^1H\}$ NMR (96 MHz, CDCl$_3$): δ −8.2 (2B), −11.2 (6B), −14.0 (2B). These data are identical with those reported in the literature [10].

1,2-[PhC=C(Ph)C(Ph)=CPh]-1,2-C$_2$B$_{10}$H$_{10}$ (V-1f). Yield: 28%. White solid. ^1H NMR (400 MHz, CDCl$_3$): δ 7.12 (m, 10H), 6.74 (m, 6H), 6.62 (m, 4H) (Ph). $^{13}C\{^1H\}$ NMR (100 MHz, CDCl$_3$): 137.8, 137.7, 137.1, 137.0, 130.9, 129.9, 127.6, 127.2, 126.8, 126.1 (Ph & olefinic C), 74.8 (cage C). $^{11}B\{^1H\}$ NMR (96 MHz, CDCl$_3$): δ −5.1 (2B), −8.8 (4B), −11.3 (4B). These data are identical with those reported in the literature [10].

1,2-[C(CH$_2$OMe)=C(CH$_2$OMe)C(CH$_2$OMe)=C(CH$_2$OMe)]-1,2-C$_2$B$_{10}$H$_{10}$ (V-1g). Yield: 13%. Colorless oil. ^1H NMR (400 MHz, CDCl$_3$): δ 4.29 (s, 4H), 4.27 (s, 4H) (OCH_2), 3.38 (s, 6H), 3.33 (s, 6H) (OCH_3). $^{13}C\{^1H\}$ NMR (100 MHz, CDCl$_3$): 136.2, 133.2 (olefinic C), 74.7 (cage C), 70.1, 67.4 (OCH$_2$), 58.3, 58.2 (OCH$_3$). $^{11}B\{^1H\}$ NMR (128 MHz, CDCl$_3$): δ −6.7 (2B), −10.6 (6B), −13.1 (2B). HRMS: m/z Calcd for $C_{14}H_{30}B_{10}O_4^+$: 370.3149. Found: 370.3145.

1,2-[MeC=C(^{i}Pr)C(Me)=C^{i}Pr]-1,2-C$_2$B$_{10}$H$_{10}$ (**V-1h**) + **1,2-[MeC=C(^{i}Pr)-C(^{i}Pr)=CMe]-1,2-C$_2$B$_{10}$H$_{10}$** (**V-1'h**).: Yield: 44%. White solid. V-**1 h**: **V-1'h** = 70:30. **V-1 h**: ^{1}H NMR (400 MHz, CDCl$_3$): δ 3.23 (m, 1H), 3.10 (m, 1H) (C*H*), 2.26 (s, 3H), 2.02 (s, 3H) (C*H*$_3$), 1.29 (d, *J* = 7.2 Hz, 6H), 1.19 (d, *J* = 7.2 Hz, 6H) (CHC*H*$_3$). ^{13}C{^{1}H} NMR (100 MHz, CDCl$_3$): δ 138.2, 129.7, 129.0 (olefinic *C*), 34.6 (C$_{cage}$-*C*–CH$_3$), 29.7, 28.5 (*C*H), 20.9, 20.7 (*C*HCH$_3$), 18.5 (C$_{cage}$-C=C–*C*H$_3$), cage *C* atoms were not observed. ^{11}B{^{1}H} NMR (128 MHz, CDCl$_3$): δ −7.6 (2B), −10.5 (6B), −13.8 (2B). HRMS: *m/z* calcd for C$_{14}$H$_{30}$B$_{10}$$^+$: 306.3345. Found: 306.3346. **V-1'h**: ^{1}H NMR (400 MHz, CDCl$_3$): δ 3.21 (m, 2H) (C*H*), 1.97 (s, 6H) (C*H*$_3$), 1.26 (d, *J* = 7.2 Hz, 12H) (CHC*H*$_3$). Compound **V-1h** was isolated as a pure product whereas **V-1'h** was always contaminated with **V-1h**. Their molar ratio was determined by ^{1}H NMR spectrum of a crude mixture.

1,2-[MeC=C(Ph)C(Me)=CPh]-1,2-C$_2$B$_{10}$H$_{10}$ (V-1i). Yield: 50%. White solid. ^{1}H NMR (400 MHz, CDCl$_3$): δ 7.38 (m, 6H), 7.16 (m, 2H), 7.02 (m, 2H) (Ph), 1.97 (s, 3H), 1.22 (s, 3H) (C*H*$_3$). ^{13}C{^{1}H} NMR (100 MHz, CDCl$_3$): 139.1, 137.8, 135.4, 133.4, 132.9, 130.2, 129.8, 128.7, 128.3, 128.1, 127.5 (Ph & olefinic *C*), 75.6 (cage *C*), 21.1, 20.6 (*C*H$_3$). ^{11}B{^{1}H} NMR (96 MHz, CDCl$_3$): δ −6.5 (2B), −10.1 (5B), −12.4 (3B). These data are identical with those reported in the literature [10].

1,2-[MeC=C(4′-Me-C$_6$H$_4$)C(Me)=C(4′-Me-C$_6$H$_4$)]-1,2-C$_2$B$_{10}$H$_{10}$ (**V-1j**). Yield: 39%. White solid. ^{1}H NMR (400 MHz, CDCl$_3$): δ 7.19 (m, 4H), 7.03 (d, *J* = 8.0 Hz, 2H), 6.90 (d, *J* = 8.0 Hz, 2H) (aromatic *H*), 2.38 (s, 3H), 2.37 (s, 3H), 1.97 (s, 3H), 1.24 (s, 3H) (C*H*$_3$). ^{13}C{^{1}H} NMR (100 MHz, CDCl$_3$): 137.8, 137.1, 136.2, 135.4, 134.9, 133.2, 132.7, 130.0, 129.4, 128.7, 128.2 (Ph & olefinic *C*), 75.8, 75.7 (cage *C*), 21.3, 21.2, 21.1, 20.7 (*C*H$_3$). ^{11}B{^{1}H} NMR (96 MHz, CDCl$_3$): δ −7.5 (2B), −10.3 (5B), −13.0 (3B). HRMS: m/z Calcd for C$_{22}$H$_{30}$B$_{10}$$^+$: 402.3345. Found: 402.3357.

1,2-[MeC=C(4′-CF$_3$-C$_6$H$_4$)C(Me)=C(4′-CF$_3$-C$_6$H$_4$)]-1,2-C$_2$B$_{10}$H$_{10}$ (**V-1k**). Yield: 49%. White solid. ^{1}H NMR (400 MHz, CDCl$_3$): δ 7.68 (m, 4H), 7.30 (d, *J* = 8.0 Hz, 2H), 7.18 (d, *J* = 8.0 Hz, 2H) (aromatic *H*), 1.97 (s, 3H), 1.20 (s, 3H) (C*H*$_3$). ^{13}C{^{1}H} NMR (100 MHz, CDCl$_3$): 142.5, 140.9, 134.3, 134.1, 132.5, 130.6, 129.5, 128.9, 126.0, 125.3 (Ph & olefinic *C*), 75.2, 74.7 (cage *C*), 21.3, 20.7 (*C*H$_3$). ^{11}B{^{1}H} NMR (128 MHz, CDCl$_3$): δ −7.4 (2B), −11.5 (5B), −13.4 (3B). HRMS: m/z Calcd for C$_{22}$H$_{24}$B$_{10}$F$_6$$^+$: 510.2780. Found: 510.2775.

1,2-[EtC=C(Ph)C(Et)=CPh]-1,2-C$_2$B$_{10}$H$_{10}$ (V-1l). Yield: 49%. White solid. ^{1}H NMR (400 MHz, CDCl$_3$): δ 7.37 (m, 6H), 7.16 (m, 4H) (Ph), 2.35 (q, *J* = 7.2 Hz, 2H), 1.65 (q, *J* = 7.2 Hz, 2H) (C*H*$_2$), 0.93 (t, *J* = 7.2 Hz, 3H), 0.48 (t, *J* = 7.2 Hz, 3H) (C*H*$_3$). ^{13}C{^{1}H} NMR (100 MHz, CDCl$_3$): 139.5, 137.7, 137.2, 135.8, 135.6, 133.8, 130.5, 129.4, 128.1, 127.7, 127.6 (Ph & olefinic *C*), 75.3, 75.2 (cage *C*), 27.4, 25.0 (*C*H$_2$), 14.4, 13.9 (*C*H$_3$). ^{11}B{^{1}H} NMR (128 MHz, CDCl$_3$): δ −7.2 (2B), −10.5 (6B), −13.1 (2B). HRMS: m/z Calcd for C$_{22}$H$_{30}$B$_{10}$$^+$: 402.3345. Found: 402.3345.

1,2-[^{n}BuC=C(Ph)C(^{n}Bu)=CPh]-1,2-C$_2$B$_{10}$H$_{10}$ (V-1m). Yield: 43%. Colorless oil. ^{1}H NMR (400 MHz, CDCl$_3$): δ 7.35 (m, 6H), 7.17 (m, 2H), 7.12 (m, 2H) (Ph), 2.27 (m, 2H), 1.58 (m, 2H), 1.32 (m, 2H), 1.06 (m, 2H), 0.88 (m, 2H) (C*H*$_2$), 0.64 (t, *J* = 7.6 Hz, 3H) (C*H*$_3$), 0.57 (m, 2H) (C*H*$_2$), 0.33 (t, *J* = 7.6 Hz,

8 Experimental Section

3H) (CH$_3$). ^{13}C{^{1}H} NMR (100 MHz, CDCl$_3$): 138.4, 137.7, 137.3, 135.7, 134.7, 133.7, 130.5, 129.5, 128.1, 128.0, 127.7, 127.5 (Ph & olefinic C), 75.5, 75.3 (cage C), 34.1, 31.8, 31.4, 31.3, 22.7, 22.3 (CH$_2$), 13.3, 13.0 (CH$_3$). ^{11}B{^{1}H} NMR (128 MHz, CDCl$_3$): δ −7.1 (2B), −10.4 (4B), −12.9 (4B). HRMS: m/z Calcd for C$_{26}$H$_{38}$B$_{10}^+$: 458.3971. Found: 458.3967.

1,2-[C(C≡CPh)=C(Ph)C(C≡CPh)=CPh]-1,2-C$_2$B$_{10}$H$_{10}$ (V-1n). Yield: 51%. Yellow solid. ^{1}H NMR (400 MHz, CDCl$_3$): δ 7.50 (m, 8H), 7.35 (m, 4H), 7.18 (m, 4H), 7.16 (m, 2H), 6.48 (d, J = 8.0 Hz, 2H) (Ph). ^{13}C{^{1}H} NMR (100 MHz, CDCl$_3$): 142.6, 140.2, 137.8, 137.6, 131.6, 131.2, 129.9, 129.5, 129.2, 128.7, 128.3, 128.0, 127.9, 127.6, 122.0, 121.9, 119.9 (Ph & olefinic C), 101.0, 100.0, 87.2, 86.3 (alkyne C), 73.9, 72.2 (cage C). ^{11}B{^{1}H} NMR (128 MHz, CDCl$_3$): δ − 7.1 (3B), −10.4 (4B), −13.0 (3B). HRMS: m/z Calcd for C$_{34}$H$_{30}$B$_{10}^+$: 546.3345. Found: 546.3336.

1,2-[C(CH$_2$OMe)=C(Ph)C(CH$_2$OMe)=CPh]-1,2-C$_2$B$_{10}$H$_{10}$ (V-1o). Yield: 24%. White solid. ^{1}H NMR (400 MHz, CDCl$_3$): δ 7.37 (m, 6H), 7.22 (m, 4H) (Ph), 3.86 (s, 2H), 3.22 (s, 2H) (OCH$_2$), 3.09 (s, 3H), 2.58 (s, 3H) (OCH$_3$). ^{13}C{^{1}H} NMR (100 MHz, CDCl$_3$): 140.9, 139.8, 136.1, 135.9, 133.1, 130.4, 130.1, 129.2, 128.5, 127.7, 127.6, 127.4 (Ph & olefinic C), 74.7, 74.6 (cage C), 70.9, 69.4 (OCH$_2$), 58.0, 57.7 (OCH$_3$). ^{11}B{^{1}H} NMR (128 MHz, CDCl$_3$): δ −6.6 (2B), −10.4 (4B), −12.7 (4B). HRMS: m/z Calcd for C$_{22}$H$_{30}$B$_{10}$O$_2^+$: 434.3254. Found: 434.3251.

1,2-[PhC=C(CH$_2$OMe)C(CH$_2$OMe)=CPh]-1,2-C$_2$B$_{10}$H$_{10}$ (V-1′o). Yield: 2%. White solid. ^{1}H NMR (400 MHz, CDCl$_3$): δ 7.41 (m, 6H), 7.25 (m, 4H) (Ph), 3.84 (s, 4H) (OCH$_2$), 3.09 (s, 6H) (OCH$_3$). ^{13}C{^{1}H} NMR (100 MHz, CDCl$_3$): 140.7, 136.1, 130.2, 130.1, 128.6, 127.7 (Ph & olefinic C), 74.2 (cage C), 69.3 (OCH$_2$), 58.2 (OCH$_3$). ^{11}B{^{1}H} NMR (128 MHz, CDCl$_3$): δ −6.3 (2B), −10.4 (4B), −12.7 (4B). HRMS: m/z Calcd for C$_{22}$H$_{30}$B$_{10}$ O$_2^+$: 434.3254. Found: 434.3249.

1-[C(CH$_2$OMe)=CH(Ph)]-1,2-C$_2$B$_{10}$H$_{11}$ (V-6o). Yield: 8%. White solid. ^{1}H NMR (400 MHz, CDCl$_3$): δ 7.34 (m, 3H), 7.22 (m, 3H) (Ph & olefinic), 4.10 (s, 1H) (cage CH), 3.96 (s, 2H) (OCH$_2$), 3.27 (s, 3H) (OCH$_3$). ^{13}C{^{1}H} NMR (100 MHz, CDCl$_3$): 140.0, 134.9, 130.4, 128.7, 128.6, 128.5 (Ph & olefinic C), 69.4 (cage C), 59.4 (OCH$_2$), 57.9 (OCH$_3$), the other cage C was not observed. ^{11}B{^{1}H} NMR (128 MHz, CDCl$_3$): δ −3.3 (2B), −10.0 (2B), −11.6 (4B), −13.7 (2B). HRMS: m/z Calcd for C$_{12}$H$_{22}$B$_{10}$O$^+$: 290.2674. Found: 290.2670.

1-[C(Ph)=CH(CH$_2$OMe)]-1,2-C$_2$B$_{10}$H$_{11}$ (V-6′o). Yield: 4%. Colorless oil. ^{1}H NMR (400 MHz, CDCl$_3$): δ 7.37 (m, 3H), 7.13 (m, 2H) (Ph), 7.01 (br s, 1H) (olefinic), 4.08 (d, J = 0.8 Hz, 2H) (OCH$_2$), 3.72 (s, 1H) (cage CH), 3.43 (s, 3H) (OCH$_3$). ^{13}C{^{1}H} NMR (100 MHz, CDCl$_3$): 135.2, 133.0, 129.9, 128.8, 128.1, 127.6 (Ph & olefinic C), 76.1, 73.9 (cage C), 58.9 (OCH$_2$), 58.1 (OCH$_3$). ^{11}B{^{1}H} NMR (128 MHz, CDCl$_3$): δ −3.3 (2B), −10.0 (2B), −11.6 (4B), −13.7 (2B). HRMS: m/z Calcd for C$_{12}$H$_{22}$B$_{10}$O$^+$: 290.2674. Found: 290.2672.

1,2-[MeC=C-(CH$_2$)$_3$-C=CMe]-1,2-C$_2$B$_{10}$H$_{10}$ (V-8a). Yield: 38%. White solid. ^{1}H NMR (400 MHz, CDCl$_3$): δ 2.49 (t, J = 7.2 Hz, 4H) (C=CCH$_2$), 2.12 (s, 6H) (CH$_3$), 1.81 (m, 2H) (CH$_2$CH$_2$CH$_2$). ^{13}C{^{1}H} NMR (100 MHz, CDCl$_3$): 136.1,

125.1 (olefinic C), 79.1 (cage C), 31.6, 23.9 (CH_2) 18.9 (CH_3). $^{11}B\{^1H\}$ NMR (96 MHz, CDCl$_3$): δ −7.5 (2B), −11.0 (5B), −12.3 (3B). HRMS: m/z Calcd for $C_{11}H_{22}B_{10}{}^+$: 262.2719. Found: 262.2710.

1,2-[MeC=C-(CH$_2$)$_4$-C=CMe]}-1,2-C$_2$B$_{10}$H$_{10}$ (V-8b). Yield: 39%. White solid. ^{1}HNMR (400 MHz, CDCl$_3$): δ 2.33 (m, 4H) (C=CCH_2), 2.12 (s, 6H) (CH_3), 1.58 (m, 4H) (CH_2CH_2CH_2CH_2). $^{13}C\{^1H\}$ NMR (100 MHz, CDCl$_3$): 128.9, 128.0 (olefinic C), 76.7 (cage C), 27.7, 21.9 (CH_2) 18.9 (CH_3). $^{11}B\{^1H\}$ NMR (96 MHz, CDCl$_3$): δ −7.0 (2B), −10.6 (4B), −12.9 (4B). HRMS: m/z Calcd for $C_{12}H_{24}B_{10}{}^+$: 276.2876. Found: 262.2868.

1,2-[MeC=C-(CH$_2$)$_5$-C=CMe]}-1,2-C$_2$B$_{10}$H$_{10}$ (V-8c). Yield: 15%. White solid. ^{1}H NMR (400 MHz, CDCl$_3$): δ 2.47 (m, 4H) (C=CCH_2), 2.20 (s, 6H) (CH_3), 1.54 (m, 6H) (CH_2). $^{13}C\{^1H\}$ NMR (100 MHz, CDCl$_3$): 134.1, 126.9 (olefinic C), 29.4, 28.6, 27.5 (CH_2), 19.4 (CH_3), cage C was not observed. $^{11}B\{^1H\}$ NMR (96 MHz, CDCl$_3$): δ −8.0 (2B), −11.1 (6B), −13.9 (2B). HRMS: m/z Calcd for $C_{13}H_{26}B_{10}{}^+$: 290.3032. Found: 290.3032.

Preparation of [{[2-C(^{n}Bu)=C(o-C$_5$H$_4$N)-1,2-C$_2$B$_{10}$H$_{10}$]Ni}$_2$(μ_2-Cl)] [Li(THF)$_4$] (V-9). To a toluene solution (5 mL) of Li$_2$C$_2$B$_{10}$H$_{10}$ (0.5 mmol), prepared in situ from the reaction of ^{n}BuLi (1.6 M, 0.63 mL, 1.0 mmol) with o-carborane (72 mg, 0.5 mmol), was added I$_2$ (127 mg, 0.5 mmol), and the reaction mixture was stirred at room temperature for 0.5 h. NiCl$_2$(PPh$_3$)$_2$ (327 mg, 0.5 mmol), and n-butyl-2-pyridinylacetylene (159 mg, 2.0 mmol) were then added and the reaction vessel was closed and heated at 90 °C overnight. Removal of the solvent gave a red residue which was washed with hexane (50 mL × 3). Recrystallization from THF at room temperature afforded **V-9**· THF as red crystals (70 mg, 25%). ^{1}H NMR (300 MHz, [D$_5$]pyridine): δ 8.16 (d, J = 4.2 Hz, 2H), 7.08 (m, 2H), 6.76 (d, J = 8.1 Hz, 2H), 6.57 (m, 2H) (Py), 3.64 (m, 10H) (CH_2O THF), 2.11 (t, J = 8.1 Hz, 4H) (CH_2), 1.59 (m, 10H) (CH_2 THF), 1.54 (m, 4H), 1.07 (m, 4H) (CH_2), 0.64 (t, J = 7.5 Hz, 6H) (CH_3). $^{13}C\{^1H\}$ NMR (100 MHz, [D$_5$]pyridine): 167.3, 158.8, 148.1, 146.8, 133.9, 121.5, 117.5 (Py & olefinic C), 90.8, 75.1 (cage C), 67.1 (CH_2O THF), 31.4, 31.0 (CH_2), 25.1 (CH_2 THF), 22.3 (CH_2), 13.2 (CH_3). $^{11}B\{^1H\}$ NMR (96 MHz, [D$_5$]pyridine): δ −3.6 (6B), −6.6 (4B), −9.5 (8B), −12.9 (2B). Anal. Calcd for $C_{46}H_{86}B_{20}ClLiN_2Ni_2O_5$ (**V-7** + THF): C, 49.19; H, 7.72; N, 2.49. Found: C, 49.07; H, 7.57; N, 2.27.

Preparation of 3-I-1,2-C$_2$B$_{10}$H$_{11}$ (VI-1a). o-Carborane (1.44 g, 10 mmol) and KOH (2.24 g, 40 mmol) were dissolved in 100 mL of MeOH and the resulting solution was refluxed for 4 h. After removal of the solvent, the residue was dissolved in 20 mL of water. Addition of a saturated Me$_3$NH·HCl aq. solution gave [Me$_3$NH][C$_2$B$_9$H$_{12}$] as a white solid (1.60 g, 83%). To a ether suspension (10 mL) of [Me$_3$NH][C$_2$B$_9$H$_{12}$] (1.60 g, 8.3 mmol) was added ^{n}BuLi (1.6 M, 1.66 mol, 1.04 mL) at 0 °C. The reaction mixture was stirred at r. t. for 2 h and then refluxed for 4 h. After removal of solvent in vacuo, hexane (20 mL) was added to the resulting solid. A solution of BI$_3$ (4.88 g, 12.5 mmol) in hexane (10 mL) was

8 Experimental Section 121

slowly added to the above suspension at 0 °C. The reaction mixture was stirred at r.t. for another 6 h and then hydrolyzed with 2 mL of water. The organic layer was separated, dried over MgSO$_4$, and concentrated. The residue was subject to column chromatography on silica gel (230–400 mesh) using hexane as eluent to give **VI-1a** (1.46 g, 65%) as a white solid. ^{1}H NMR (400 MHz, CDCl$_3$): δ 3.88 (s, 2H) (cage *H*). These data are identical with those reported in the literature [6].

Preparation of 3-I-1-Me-1,2-C$_2$B$_{10}$H$_{10}$ (VI-1b). To an Et$_2$O solution (20 mL) of *o*-carborane (1.44 g, 10 mmol) was added dropwise ^{n}BuLi (1.6 M, 6.25 mL, 10 mmol) at 0 °C, and the mixture was stirred at r. t. for 1 h. To the resulting reaction mixture was added TMSCl (1.27 mL, 10 mmol) at 0 °C, and the solution was stirred at room temperature for another 4 h. ^{n}BuLi (1.6 M, 6.25 mL, 10 mmol) was added at 0 °C, and the reaction mixture was stirred at room temperature for 1 h before 2 equiv of MeI (1.25 mL, 20 mmol) was added at 0 °C. After stirring at room temperature overnight, the reaction was quenched with 20 mL of water and extracted with Et$_2$O (10 mL × 3). After removal of the solvents in vacuo, 1-methyl-2-trimethylsilyl-*o*-carborane was obtained as a white solid, which was used in the next step reaction without further purification (2.28 g, 99%). 1-Methyl-2-trimethylsilyl-*o*-carborane (2.28 g, 9.9 mmol) and KOH (2.24 g, 40 mmol) were dissolved in 100 mL of MeOH and the resulting solution was refluxed for 4 h. After removal of the solvent, the residue was dissolved in 20 mL of water. Addition of a saturated Me$_3$NH·HCl aq. solution gave [Me$_3$NH][3-Me-7,8-C$_2$B$_9$H$_{11}$] as a white solid (1.82 g, 88%). To a ether suspension (10 mL) of [Me$_3$NH][3-Me-7,8-C$_2$B$_9$H$_{11}$] (1.82 g, 8.7 mmol) was added ^{n}BuLi (1.6 M, 1.74 mol, 1.09 mL) at 0 °C. The reaction mixture was stirred at r. t. for 2 h and then refluxed for 4 h. After removal of solvent in vacuo, hexane (20 mL) was added to the resulting solid. A solution of BI$_3$ (5.12 g, 13.1 mmol) in hexane (10 mL) was slowly added to the above suspension at 0 °C. The reaction mixture was stirred at r.t. for another 6 h and then hydrolyzed with 2 mL of water. The organic layer was separated, dried over MgSO$_4$, and concentrated. The residue was subject to column chromatography on silica gel (230–400 mesh) using hexane as eluent to give **VI-1b** (1.51 g, 61%) as a white solid. ^{1}H NMR (400 MHz, CDCl$_3$): δ 3.61 (s, 1H) (cage *H*), 2.24 (s, 3H) (C*H$_3$*). These data are identical with those reported in the literature [13].

Preparation of 3-I-1-Ph-1,2-C$_2$B$_{10}$H$_{10}$ (VI-1c). **VI-1c** (640 mg, 37%) was prepared as a white solid from 1-phenyl-*o*-carborane (1.10 g, 5 mmol) using the same method for **VI-1a**. ^{1}H NMR (400 MHz, CDCl$_3$): δ 7.44 (m, 3H), 7.34 (m, 2H) (Ph), 4.24 (s, 1H) (cage *H*). These data are identical with those reported in the literature [13].

Preparation of 1-^{n}Bu-3-I-1,2-C$_2$B$_{10}$H$_{10}$ (VI-1d). To an ether solution (5 mL) of 3-iodo-*o*-carborane (400 mg, 1.5 mmol) was added ^{n}BuLi (1.6 M, 1.5 mmol, 0.93 mL) and the mixture was stirred at 0 °C for 1 h. After adding ^{n}BuBr (1.5 mmol, 0.16 mL), the reaction mixture was stirred for at 0 °C 5 h and then hydrolyzed with water. The organic layer was separated, dried over MgSO$_4$, and concentrated. The residue was subject to column chromatography on silica gel (230–400 mesh) using hexane as eluent to give **VI-1d** (254 g, 53%) as a colorless oil. ^{1}H NMR (400 MHz, CDCl$_3$): δ 3.61 (s, 1H) (cage H), 2.47 (m, 1H), 2.30 (m, 1H), 1.50 (m, 2H), 1.35 (m, 2H) (CH_2), 0.94 (t, $J = 7.2$ Hz, 3H) (CH_3). ^{13}C$\{^1$H$\}$ NMR (100 MHz, CDCl$_3$): δ 75.0, 64.7 (cage C), 38.9, 30.9, 22.0 (CH_2), 13.6 (CH_3). ^{11}B$\{^1$H$\}$ NMR (128 MHz, CDCl$_3$): δ -2.6 (1B), -5.2 (1B), -7.5 (1B), -10.0 (1B), -11.3 (4B), -13.3 (1B), -25.3 (1B). HRMS: m/z calcd for C$_6$H$_{19}$B$_{10}$I$^+$: 326.1529. Found: 326.1532.

Preparation of 3-I-1-TMS-1,2-C$_2$B$_{10}$H$_{10}$ (VI-1e). VI-1e was prepared as a white solid from TMSCl using the same method for **VI-1d**. Yield: 87%. ^{1}H NMR (400 MHz, CDCl$_3$): δ 3.64 (s, 1H) (cage H), 0.41 (s, 9H) (CH_3). ^{13}C$\{^1$H$\}$ NMR (100 MHz, CDCl$_3$): δ 67.4, 64.1 (cage C), -0.9 (CH_3). ^{11}B$\{^1$H$\}$ NMR (128 MHz, CDCl$_3$): δ -0.2 (1B), -1.7 (1B), -4.7 (1B), -9.0 (2B), -9.7 (1B), -11.0 (1B), -12.1 (1B), -13.0 (1B), -28.1 (1B). HRMS: m/z calcd for C$_5$H$_{20}$B$_{10}$ISi [M-H]$^+$: 342.1298. Found: 342.1302.

Preparation of 3-I-1-(CH$_2$CH$_2$OCH$_3$)-1,2-C$_2$B$_{10}$H$_{10}$ (VI-1f). VI-1f was prepared as a colorless oil from 2-chloroethyl methyl ether using the same method for **VI-1d**. Yield: 85%. ^{1}H NMR (400 MHz, CDCl$_3$): δ 3.87 (s, 1H) (cage H), 3.56 (m, 2H) (OCH_2), 3.32 (s, 3H) (OCH_3), 2.79 (m, 1H), 2.63 (m, 1H) (OCH$_2$CH_2). ^{13}C$\{^1$H$\}$ NMR (100 MHz, CDCl$_3$): δ 72.8, 64.4 (cage C), 70.0 (OCH_2), 58.7 (OCH_3), 38.3 (OCH$_2$CH_2). ^{11}B$\{^1$H$\}$ NMR (128 MHz, CDCl$_3$): δ -2.2 (1B), -4.7 (1B), -7.7 (1B), -9.7 (1B), -11.3 (4B), -12.7 (1B), -24.5 (1B). HRMS: m/z calcd for C$_5$H$_{17}$B$_{10}$IO$^+$: 328.1322. Found: 328.1323.

Preparation of 3-I-1-[CH$_2$CH$_2$N(CH$_3$)$_2$]-1,2-C$_2$B$_{10}$H$_{10}$ (VI-1 g). VI-1g was prepared as a light yellow oil from 2 equiv ^{n}BuLi and 2-chloro-N,N-dimethyl-ethylamine hydrochloride using the same method for **VI-1d**. Yield: 66%. ^{1}H NMR (400 MHz, CDCl$_3$): δ 4.00 (s, 1H) (cage H), 2.59 (m, 3H), 2.41 (m, 1H) (CH_2), 2.23 (s, 6H) (NCH_3). ^{13}C$\{^1$H$\}$ NMR (100 MHz, CDCl$_3$): δ 73.7, 64.4 (cage C), 57.8 (NCH_2), 45.4 (NCH_3), 35.8 (CH_2CH$_2$N). ^{11}B$\{^1$H$\}$ NMR (96 MHz, CDCl$_3$): δ -1.9 (1B), -4.3 (1B), -7.4 (1B), -9.5 (1B), -11.0 (4B), -12.5 (1B), -24.2 (1B). HRMS: m/z calcd for C$_6$H$_{20}$B$_{10}$IN$^+$: 341.1638. Found: 341.1639.

General Procedure for Palladium/Nickel-Cocatalyzed Cycloaddition Reaction of 1,3-o-Carboryne with Alkynes. To a toluene solution (5 mL) of 3-iodo-1-methyl-o-carborane (0.5 mmol) was added 1 equiv of ^{n}BuLi (0.5 mmol), and the reaction mixture was stirred at room temperature for 0.5 h. Pd(PPh$_3$)$_4$ (5 mol%), Ni(cod)$_2$ (5 mol%), and alkyne (2.0 mmol) [or diyne (1.0 mmol)] were then added, and the reaction vessel was closed and heated at 110 °C overnight. After addition of water and extraction with ether, the resulting solution was concentrated to dryness in vacuo. The residue was subject to column chroma-tography on silica gel (230–400 mesh) using hexane as eluent to give the desired cycloaddition product.

1,3-[EtC=C(Et)C(Et)=CEt]-1,2-C$_2$B$_{10}$H$_{10}$ (VI-4a). Yield: 12%. Colorless crystals. ^{1}H NMR (400 MHz, CDCl$_3$): δ 2.61 (m, 4H), 2.39 (m, 5H) (cage CH & CH_2), 1.19 (t, $J = 7.2$ Hz, 3H), 1.10 (t, $J = 7.2$ Hz, 3H), 1.03 (t, $J = 7.2$ Hz, 3H), 1.01 (t, $J = 7.2$ Hz, 3H), (CH_3). ^{1}H NMR (400 MHz, benzene-d_6): δ 2.54 (q, $J = 7.6$ Hz, 2H), 2.23 (m, 2H), 2,15 (m, 1H), 2.02 (m, 3H) (CH_2), 1.83 (s, 1H) (cage CH), 1.15 (t, $J = 7.6$ Hz, 3H), 0.87 (t, $J = 7.6$ Hz, 3H), 0.82 (t, $J = 7.6$ Hz,

8 Experimental Section   123

3H), 0.74 (t, $J = 7.6$ Hz, 3H), (CH$_3$). ^{13}C{^{1}H} NMR (75 MHz, CDCl$_3$): δ 143.3, 142.0, 130.6 (olefinic *C*), 60.4 (cage *C*), 28.3, 27.0, 23.3, 21.9 (CH$_2$), 15.5, 15.1, 15.0, 14.8 (CH$_3$), the olefinic *C* connected to B atom and another cage *C* were not observed. ^{11}B{^{1}H} NMR (96 MHz, CDCl$_3$): δ −4.8 (1B), −7.8 (2B), −10.0 (1B), −11.7 (3B), −13.2 (1B), −14.0 (1B), −16.8 (1B). HRMS: *m/z* calcd for C$_{14}$H$_{30}$B$_{10}$$^+$: 306.3345. Found: 306.3349.

2-Me-1,3-[EtC=C(Et)C(Et)=CEt]-1,2-C$_2$B$_{10}$H$_9$ (VI-4b). Yield: 54%. Colorless crystals. ^{1}H NMR (400 MHz, CDCl$_3$): δ 2.54 (m, 3H), 2.44 (m, 5H) (CH$_2$), 1.29 (s, 3H) (CH$_3$), 1.15 (t, $J = 7.6$ Hz, 3H), 1.12 (t, $J = 7.6$ Hz, 3H), 1.05 (t, $J = 7.6$ Hz, 3H), 1.04 (t, $J = 7.6$ Hz, 3H) (CH$_2$CH$_3$). ^{13}C{^{1}H} NMR (100 MHz, CDCl$_3$): δ 145.6, 143.5, 128.8 (olefinic *C*), 81.3, 67.9 (cage *C*), 28.2, 26.9, 23.5, 21.9 (CH$_2$), 20.2 (CH$_3$), 15.6, 15.3, 15.2, 14.7 (CH$_2$CH$_3$), the olefinic *C* connected to B atom was not observed. ^{11}B{^{1}H} NMR (96 MHz, CDCl$_3$): δ −8.0 (3B), −9.9 (1B), −11.2 (3B), −12.8 (1B), −14.4 (2B). HRMS: *m/z* calcd for C$_{15}$H$_{32}$B$_{10}$$^+$: 320.3502. Found: 320.3504.

2-^{n}Bu-1,3-[EtC=C(Et)C(Et)=CEt]-1,2-C$_2$B$_{10}$H$_9$ (VI-4c). Yield: 67%. Colorless oil. ^{1}H NMR (400 MHz, CDCl$_3$): δ 2.43 (m, 10H), 1.33 (m, 2H), 1.25 (m, 2H) (CH$_2$), 1.15 (t, $J = 7.2$ Hz, 3H), 1.11 (t, $J = 7.2$ Hz, 3H), 1.04 (t, $J = 7.2$ Hz, 3H), 1.02 (t, $J = 7.2$ Hz, 3H), 0.79 (t, $J = 7.2$ Hz, 3H) (CH$_3$). ^{13}C{^{1}H} NMR (100 MHz, CDCl$_3$): δ 145.8, 143.3, 128.7 (olefinic *C*), 82.9, 72.7 (cage *C*), 31.4, 31.2, 28.1, 27.0, 23.5, 22.4, 21.8 (CH$_2$), 15.4, 15.3, 15.2, 14.8, 13.6 (CH$_3$), the olefinic *C* connected to B atom was not observed. ^{11}B{^{1}H} NMR (128 MHz, CDCl$_3$): δ −7.8 (3B), −11.2 (5B), −14.4 (1B), −16.2 (1B). HRMS: *m/z* calcd for C$_{18}$H$_{38}$B$_{10}$$^+$: 362.3971. Found: 392.3967.

2-TMS-1,3-[EtC=C(Et)C(Et)=CEt]-1,2-C$_2$B$_{10}$H$_9$ (VI-4d). Yield: 69%. Colorless crystals. ^{1}H NMR (400 MHz, CDCl$_3$): δ 2.46 (m, 8H) (CH$_2$), 1.26 (t, $J = 7.6$ Hz, 3H), 1.18 (t, $J = 7.6$ Hz, 3H), 1.08 (t, $J = 7.6$ Hz, 3H), 1.06 (t, $J = 7.6$ Hz, 3H) (CH$_2$CH$_3$), 0.03 (s, 9H) (Si(CH$_3$)$_3$). ^{13}C{^{1}H} NMR (100 MHz, CDCl$_3$): δ 144.4, 142.2, 131.5 (olefinic *C*), 83.8, 68.5 (cage *C*), 29.0, 27.8, 23.5, 21.8 (CH$_2$), 15.2, 15.0, 14.7, 14.6 (CH$_2$CH$_3$), 0.56 (Si(CH$_3$)$_3$), the olefinic *C* connected to B atom was not observed. ^{11}B{^{1}H} NMR (128 MHz, CDCl$_3$): δ −2.6 (1B), −7.2 (2B), −7.8 (2B), −11.1 (3B), −12.3 (1B), −14.3 (1B). HRMS: *m/z* calcd for C$_{17}$H$_{38}$B$_{10}$Si$^+$: 378.3740. Found: 378.3748.

2-Ph-1,3-[EtC=C(Et)C(Et)=CEt]-1,2-C$_2$B$_{10}$H$_9$ (VI-4e). Yield: 43%. Colorless crystals. ^{1}H NMR (400 MHz, CDCl$_3$): δ 7.24 (m, 3H), 7.14 (m, 2H) (Ph), 2.63 (m, 4H), 2.13 (m, 3H), 1.98 (m, 1H) (CH$_2$), 1.27 (t, $J = 7.6$ Hz, 3H), 1.15 (t, $J = 7.6$ Hz, 3H), 0.64 (t, $J = 7.6$ Hz, 3H), 0.49 (t, $J = 7.6$ Hz, 3H) (CH$_3$). ^{13}C{^{1}H} NMR (100 MHz, CDCl$_3$): δ 145.3, 143.4, 130.3, 129.9, 129.4, 128.8, 127.4 (olefinic *C* & Ph), 84.9, 75.5 (cage *C*), 28.5, 27.5, 23.1, 21.6 (CH$_2$), 15.0, 14.7, 14.6, 14.0 (CH$_3$), the olefinic *C* connected to B atom was not observed. ^{11}B{^{1}H} NMR (128 MHz, CDCl$_3$): δ −6.0 (2B), −7.6 (1B), −10.0 (1B), −11.4 (3B), −13.7 (3B). HRMS: *m/z* calcd for C$_{20}$H$_{34}$B$_{10}$$^+$: 382.3658. Found: 382.3657.

2-(CH$_2$CH$_2$OCH$_3$)-1,3-[EtC=C(Et)C(Et)=CEt]-1,2-C$_2$B$_{10}$H$_9$ (VI-4f). Yield: 58%. Colorless oil. ^{1}H NMR (400 MHz, CDCl$_3$): δ 3.25 (t, $J = 7.2$ Hz, 2H) (OCH$_2$), 3.23 (s, 3H) (OCH$_3$), 2.52 (m, 3H), 2.38 (m, 5H), 1.64 (m, 2H) (CH$_2$),

1.16 (t, $J = 7.6$ Hz, 3H), 1.12 (t, $J = 7.6$ Hz, 3H), 1.05 (t, $J = 7.6$ Hz, 3H), 1.03 (t, $J = 7.6$ Hz, 3H) (CH_3). $^{13}C\{^1H\}$ NMR (100 MHz, CDCl$_3$): δ 146.0, 143.6, 128.7 (olefinic C), 82.8, 69.5 (cage C), 70.7 (OCH_2), 58.5 (OCH_3), 31.1, 28.2, 27.0, 23.5, 21.9 (CH_2), 15.4, 15.2, 15.1, 14.7 (CH_3), the olefinic C connected to B atom was not observed. $^{11}B\{^1H\}$ NMR (128 MHz, CDCl$_3$): δ −7.1 (1B), −7.8 (1B), −11.1 (5B), −12.7 (1B), −14.2 (1B), −16.1 (1B). HRMS: m/z calcd for $C_{17}H_{36}B_{10}O^+$: 364.3764. Found: 364.3760.

2-[CH$_2$CH$_2$N(CH$_3$)$_2$]-1,3-[EtC=C(Et)C(Et)=CEt]-1,2-C$_2$B$_{10}$H$_9$ (**VI-4g**). Yield: 51%. Colorless oil. ^1H NMR (400 MHz, CDCl$_3$): δ 2.44 (m, 10H) (CH_2), 2.10 (s, 6H) (N(CH_3)$_2$), 1.54 (m, 2H) (CH_2), 1.16 (t, $J = 7.6$ Hz, 3H), 1.14 (t, $J = 7.6$ Hz, 3H), 1.08 (t, $J = 7.6$ Hz, 3H), 1.04 (t, $J = 7.6$ Hz, 3H) (CH_3). $^{13}C\{^1H\}$ NMR (100 MHz, CDCl$_3$): δ 146.0, 143.6, 128.7 (olefinic C), 83.1, 70.5 (cage C), 58.4 (NCH_2), 45.3 (N(CH_3)$_2$), 29.1, 28.2, 27.0, 23.5, 21.9 (CH_2), 15.5, 15.4, 15.3, 14.9 (CH_3), the olefinic C connected to B atom was not observed. $^{11}B\{^1H\}$ NMR (128 MHz, CDCl$_3$): δ −7.6 (3B), −10.9 (5B), −14.1 (1B), −16.2 (1B). HRMS: m/z calcd for $C_{18}H_{39}B_{10}N$: 377.4080. Found: 377.4076.

2-Me-1,3-[^{n}PrC=C(^{n}Pr)C(^{n}Pr)=C^{n}Pr]-1,2-C$_2$B$_{10}$H$_9$ (**VI-4h**). Yield: 55%. White solid. ^1H NMR (400 MHz, CDCl$_3$): δ 2.48 (m, 3H), 2.27 (m, 5H) (=CCH_2), 1.51 (m, 4H), 1.36 (m, 4H) (CH$_2$$CH_2$), 1.26 (s, 3H) ($CH_3$), 0.98 (m, 12H) (CH$_2$$CH_3$). $^{13}C\{^1H\}$ NMR (100 MHz, CDCl$_3$): δ 144.4, 142.4, 127.8 (olefinic C), 81.3, 68.0 (cage C), 37.9, 36.5, 33.2, 31.6 (=CCH_2), 24.6, 24.2, 23.6 (CH_2CH_2), 20.2 (CH_3), 14.8, 14.7, 14.6, 14.4 (CH$_2$$CH_3$), the olefinic C connected to B atom was not observed. $^{11}B\{^1H\}$ NMR (96 MHz, CDCl$_3$): δ −8.3 (3B), −11.4 (5B), −14.6 (2B). HRMS: m/z calcd for $C_{19}H_{40}B_{10}^+$: 376.4128. Found: 376.4114.

2-Me-1,3-[^{n}BuC=C(^{n}Bu)C(^{n}Bu)=C^{n}Bu]-1,2-C$_2$B$_{10}$H$_9$ (**VI-4i**). Yield: 33%. The reaction of 5-nonyne catalyzed by 10 mol % Pd(PPh$_3$)$_4$ or 10 mol % [Pd(Ally)Cl]$_2$/20 mol % PPh$_3$ was completed in about 7 days and 5 days to give **VII-4i** in 26 and 23% yields, respectively. Colorless oil. ^1H NMR (400 MHz, CDCl$_3$): δ 2.48 (m, 5H), 2.31 (m, 7H), 1.40 (m, 12H) (CH_2), 1.27 (s, 3H) (CH_3), 0.95 (m, 12H) (CH$_2$$CH_3$). $^{13}C\{^1H\}$ NMR (75 MHz, CDCl$_3$): δ 144.4, 142.3, 127.6 (olefinic C), 81.4, 68.0 (cage C), 35.4, 34.0, 33.4, 33.2, 32.9, 32.4 30.7, 29.1, 23.4, 23.3, 23.2, 23.1(CH_2), 20.2 (CH_3), 14.0, 13.9, 13.8, 13.7 (CH$_2$$CH_3$), the olefinic C connected to B atom was not observed. $^{11}B\{^1H\}$ NMR (96 MHz, CDCl$_3$): δ −8.4 (4B), −11.5 (4B), −14.6 (2B). HRMS: m/z calcd for $C_{23}H_{48}B_{10}^+$: 432.4754. Found: 432.4758.

2-Me-1,3-[PhC=C(Ph)C(Ph)=CPh]-1,2-C$_2$B$_{10}$H$_9$ (**VI-4j**). Yield: 55%. Yellow crystals. ^1H NMR (400 MHz, CDCl$_3$): δ 7.09 (m, 9H), 6.89 (d, $J = 8.0$ Hz, 1H), 6.77 (m, 10H) (aromatic CH), 2.12 (s, 3H) (CH_3). $^{13}C\{^1H\}$ NMR (100 MHz, CDCl$_3$): δ 147.5, 145.4, 142.2, 139.7, 139.5, 139.0, 131.9, 131.5, 130.4, 130.1, 129.7, 128.7, 127.5, 127.3, 127.2, 127.1, 126.8, 126.6, 126.5, 126.0, 125.9, 125.7 (aromatic & olefinic C), 78.7, 67.8 (cage C), 21.0 (CH_3), the olefinic C connected to B atom was not observed. $^{11}B\{^1H\}$ NMR (96 MHz, CDCl$_3$): δ −8.2 (3B), −10.5 (4B), −13.6 (3B). HRMS: m/z calcd for $C_{31}H_{32}B_{10}^+$: 512.3502. Found: 512.3520.

2-Me-1,3-[C(4′-Me-C$_6$H$_4$)=C(4′-Me-C$_6$H$_4$)C(4′-Me-C$_6$H$_4$)=C(4′-Me-C$_6$H$_4$)]-1, 2-C$_2$B$_{10}$H$_9$ (**VI-4k**). Yield: 51%. White solid. ^1H NMR (400 MHz, CDCl$_3$):

8 Experimental Section

δ 6.90 (m, 6H), 6.83 (d, $J = 7.6$ Hz, 1H), 6.73 (d, $J = 7.6$ Hz, 1H), 6.50 (m, 8H) (aromatic C*H*), 2.23 (s, 3H), 2.21 (s, 3H), 2.03 (s, 6H), 2.00 (s, 3H) (C*H*$_3$). ^{13}C{^{1}H} NMR (100 MHz, CDCl$_3$): δ 147.7, 145.5, 139.6, 137.0, 136.8, 136.5, 136.4, 135.1, 135.0, 134.8, 131.6, 131.3, 130.2, 129.9, 129.5, 128.6, 128.2, 127.9, 127.7, 127.5, 127.2 (aromatic & olefinic *C*), 79.1, 67.9 (cage *C*), 21.0, 20.9 (*C*H$_3$), the olefinic *C* connected to B atom was not observed. ^{11}B{^{1}H} NMR (128 MHz, CDCl$_3$): δ −7.8 (4B), −10.1 (4B), −13.3 (2B). HRMS: *m/z* calcd for C$_{35}$H$_{40}$B$_{10}^+$: 568.4128. Found: 568.4150.

2-Me-1,3-[PhC=C(Me)C(Ph)=CMe]-1,2-C$_2$B$_{10}$H$_9$ **(VI-4 l) + 2-Me-1,3-[MeC=C(Ph)C(Ph)=CMe]-1,2-C$_2$B$_{10}$H$_9$ (VI-5l). Yield: 49%. White solid. VI-4l: VI-5l** = 62:38. **VI-4l:** ^{1}H NMR (400 MHz, CDCl$_3$): δ 7.36 (m, 6H), 7.12 (m, 2H), 6.93 (m, 2H) (aromatic C*H*), 1.93 (s, 3H), 1.70 (s, 3H), 1.34 (s, 3H) (C*H*$_3$). ^{13}C{^{1}H} NMR (75 MHz, CDCl$_3$): δ 144.8, 140.8, 139.8, 130.8, 129.6, 128.7, 128.5, 128.2, 128.1, 127.7, 126.9 (aromatic & olefinic *C*), 79.8, 67.9 (cage *C*), 22.9, 21.7, 20.5 (*C*H$_3$), the olefinic *C* connected to B atom was not observed. ^{11}B{^{1}H} NMR (96 MHz, CDCl$_3$): δ −7.3 (3B), −10.4 (5B), −13.2 (2B). HRMS: *m/z* calcd for C$_{21}$H$_{28}$B$_{10}^+$: 388.3189. Found: 388.3189. Compound **VI-4l** was isolated as a pure product whereas **VI-5l** was always contaminated with **VI-4l**. Their molar ratio was determined by ^{1}H NMR spectrum of a crude mixture.

2-Me-1,3-[PhC=C(Et)C(Ph)=CEt]-1,2-C$_2$B$_{10}$H$_9$ **(VI-4m) + 2-Me-1,3-[EtC=C(Ph)C(Ph)=CEt]-1,2-C$_2$B$_{10}$H$_9$ (VI-5m). Yield: 47%. White solid. VI-4m: VI-5m** = 80:20. **VII-4m:** ^{1}H NMR (400 MHz, CDCl$_3$): δ 7.32 (m, 6H), 7.20 (d, $J = 7.2$ Hz, 1H), 7.16 (m, 1H), 7.06 (m, 1H), 6.97 (m, 1H) (aromatic C*H*), 2.27 (m, 2H), 1.78 (m, 2H) (C*H*$_2$), 1.73 (s, 3H) (C*H*$_3$), 1.01 (t, $J = 7.6$ Hz, 3H), 0.55 (t, $J = 7.6$ Hz, 3H) (CH$_2$C*H*$_3$). ^{13}C{^{1}H} NMR (100 MHz, CDCl$_3$): δ 146.9, 144.7, 139.6, 139.2, 131.1, 129.9, 129.4, 128.0, 127.9, 127.8, 127.7, 127.6, 127.0 (aromatic & olefinic *C*), 79.7, 67.5 (cage *C*), 28.4, 26.8 (*C*H$_2$), 20.6 (*C*H$_3$), 14.3 (CH$_2$*C*H$_3$), the olefinic *C* connected to B atom was not observed. ^{11}B{^{1}H} NMR (96 MHz, CDCl$_3$): δ −7.7 (3B), −10.9 (4B), −13.6 (3B). HRMS: *m/z* calcd for C$_{23}$H$_{32}$B$_{10}^+$: 416.3502. Found: 416.3489. **VI-5m:** ^{1}H NMR (400 MHz, CDCl$_3$): δ 7.03 (m, 6H), 6.86 (d, $J = 7.6$ Hz, 1H), 6.81 (d, $J = 6.4$ Hz, 1H), 6.73 (d, $J = 8.0$ Hz, 1H), 6.61 (m, 1H) (aromatic C*H*), 2.23 (m, 4H) (C*H*$_2$), 1.71 (s, 3H) (C*H*$_3$), 0.98 (t, $J = 7.6$ Hz, 3H), 0.81 (t, $J = 7.6$ Hz, 3H) (CH$_2$C*H*$_3$). ^{13}C{^{1}H} NMR (100 MHz, CDCl$_3$): δ 146.1, 144.8, 140.0, 130.6, 129.7, 129.6, 129.5, 129.1, 127.4, 127.3, 127.2, 126.4, 126.0 (aromatic & olefinic *C*), 80.7, 68.0 (cage *C*), 29.0, 28.2 (*C*H$_2$), 20.5 (*C*H$_3$), 14.5, 14.4 (CH$_2$*C*H$_3$), the olefinic *C* connected to B atom was not observed. ^{11}B{^{1}H} NMR (96 MHz, CDCl$_3$): δ −7.4 (3B), −10.6 (4B), −13.6 (3B). HRMS: *m/z* calcd for C$_{23}$H$_{32}$B$_{10}^+$: 416.3502. Found: 416.3506.

2-Me-1,3-[MeC=C-(CH$_2$)$_3$-C=CMe]-1,2-C$_2$B$_{10}$H$_9$ (VI-7a). Yield: 6%. White solid. ^{1}H NMR (400 MHz, CDCl$_3$): δ 2.57 (m, 4H) (=CC*H*$_2$), 2.09 (s, 3H), 2.02 (s, 3H) (=CC*H*$_3$), 1.83 (m, 2H) (CH$_2$C*H*$_2$), 1.24 (s, 3H) (C*H*$_3$). ^{13}C{^{1}H} NMR (75 MHz, CDCl$_3$): δ 147.2, 145.3, 118.7 (olefinic *C*), 81.7, 67.7 (cage *C*), 33.1, 31.3 (=C*C*H$_2$), 23.8 (*C*H$_2$CH$_2$), 20.3, 20.1, 19.5 (*C*H$_3$), the olefinic *C* connected to

B atom was not observed. $^{11}B\{^1H\}$ NMR (96 MHz, $CDCl_3$): δ −6.2 (1B), −7.3 (2B), −9.6 (1B), −11.0 (3B), −12.6 (1B), −13.8 (2B). HRMS: m/z calcd for $[C_{12}H_{24}B_{10}]^+$: 276.2876. Found: 276.2867.

2-Me-1,3-[MeC=C-(CH₂)₄-C=CMe]-1,2-C₂B₁₀H₉ (VI-7b). Yield: 34%. Colorless crystals. ^1H NMR (400 MHz, $CDCl_3$): δ 2.45 (m, 4H) (=CCH_2), 2.11 (s, 3H), 2.04 (s, 3H) (=CCH_3), 1.65 (m, 4H) (CH₂CH_2), 1.23 (s, 3H) (CH_3). $^{13}C\{^1H\}$ NMR (75 MHz, $CDCl_3$): δ 140.8, 139.0, 121.3 (olefinic C), 81.1, 67.4 (cage C), 29.3, 27.2 (=CCH_2), 22.1, 21.9 (CH₂CH₂), 21.0, 20.1, 19.3 (CH_3), the olefinic C connected to B atom was not observed. $^{11}B\{^1H\}$ NMR (96 MHz, $CDCl_3$): δ −7.5 (3B), −9.3 (1B), −10.9 (3B), −12.2 (1B), −14.0 (2B). HRMS: m/z calcd for $C_{13}H_{26}B_{10}{}^+$: 290.3032. Found: 290.3029.

2-Me-1,3-[MeC=C-(CH₂)₅-C=CMe]-1,2-C₂B₁₀H₉ (VI-7c). Yield: 23%. White solid. ^1H NMR (400 MHz, $CDCl_3$): δ 2.61 (m, 4H) (=CCH_2), 2.16 (s, 3H), 2.11 (s, 3H) (=CCH_3), 1.67 (m, 1H), 1.58 (m, 4H), 1.48 (m, 1H) (CH₂CH_2), 1.30 (s, 3H) (CH_3). $^{13}C\{^1H\}$ NMR (75 MHz, $CDCl_3$): δ 146.0, 144.0, 124.4 (olefinic C), 81.1, 68.3 (cage C), 31.0, 29.0 (=CCH_2), 28.3, 28.0 (CH₂CH₂), 21.6, 20.2, 19.9 (CH_3), the olefinic C connected to B atom was not observed. $^{11}B\{^1H\}$ NMR (96 MHz, $CDCl_3$): δ −7.9 (3B), −9.9 (1B), −11.2 (3B), −12.6 (1B), −14.4 (2B). HRMS: m/z calcd for $C_{14}H_{28}B_{10}{}^+$: 304.3189. Found: 304.3179.

X-ray Structure Determination. Single-crystals were immersed in Paraton-N oil and sealed under N_2 in thin-walled glass capillaries. All data were collected at 293 K on a Bruker SMART 1,000 CCD diffractometer using Mo-Kα radiation. An empirical absorption correction was applied using the SADABS program [14]. All structures were solved by direct methods and subsequent Fourier difference techniques and refined anisotropically for all non-hydrogen atoms by full-matrix least squares calculations on F^2 using the SHELXTL program package [15]. All hydrogen atoms were geometrically fixed using the riding model. Crystal data and details of data collection and structure refinements are given in Appendix II. CIF files are given in Appendix III in electronic format.

References

1. Venanzi LMJ (1958) Inorg Nucl Chem 8:137
2. Herrmann WA, Gerstberger G, Spiegler M (1997) Organometallics 16:2209
3. Peña D, Pérez D, Guitián E, Castedo L (1999) Org Lett 1:1555
4. Peña D, Pérez D, Guitián E, Castedo LJ (1999) Am Chem Soc 121:5827
5. Li J, Jones M Jr (1990) Inorg Chem 29:4162
6. Li J, Logan CF, Jones M Jr (1991) Inorg Chem 30:4868
7. Andrew JS, Zayas J, Jones M Jr (1985) Inorg Chem 24:3715
8. Barberà G, Vaca A, Teixidor F, Sillanpää R, Kivekäs R, Viñas C (2008) Inorg Chem 47:7309
9. Ohta K, Goto T, Endo Y (2005) Inorg Chem 44:8569
10. Ren S, Chan H-S, Xie Z (2009) J Am Chem Soc 131:3862

References

11. Huang Q, Gingrich HL, Jones M Jr (1991) Inorg Chem 30:3254
12. Shen H, Chan H-S, Xie Z (2006) Organometallics 25:2617
13. Clara Vinas C, Barberà G, Oliva JM, Teixidor F, Welch AJ, Rosair GM (2001) Inorg Chem 40:6555
14. Sheldrick GM (1996) SADABS: program for empirical absorption correction of area detector data. University of Göttingen, Germany
15. Sheldrick GM (1997) SHELXTL 5.10 for windows NT structure determination software programs. Bruker Analytical X-ray Systems, Inc, Madison, Wisconsin, USA

Appendix

Crystal Data and Summary of Data Collection and Refinement

	II-2	II-3	II-4	II-5
Formular	$C_{38}H_{38}B_{10}I_2NiP_2$	$C_8H_{27}B_{10}BrNiP_2$	$C_{14}H_{32}B_{10}NiP_2$	$C_{44}H_{44}B_{10}NiP_2$
Crystal size (mm)	$0.40 \times 0.30 \times 0.20$	$0.40 \times 0.30 \times 0.20$	$0.40 \times 0.30 \times 0.20$	$0.50 \times 0.40 \times 0.20$
fw	977.23	431.96	429.15	801.54
Crystal system	Triclinic	Orthorhombic	Monoclinic	Monoclinic
Space group	$P(-1)$	Ama2	$P2_1$	$P2_1/n$
a, Å	12.597(2)	15.253(2)	8.937(1)	12.958(1)
b, Å	12.615(2)	11.484(2)	15.886(1)	21.107(1)
c, Å	14.992(2)	11.858(2)	17.545(2)	15.398(1)
α, deg	71.936(3)	90	90	90
β, deg	78.406(3)	90	104.205(2)	92.682(1)
γ, deg	71.484(2)	90	90	90
V, Å^3	2133.9(5)	2077.0(5)	2414.9(4)	4188.9(4)
Z	2	4	4	4
D_{calcd}, Mg/m^3	1.521	1.381	1.180	1.271
Radiation (λ), Å	0.71073	0.71073	0.71073	0.71073
2θ range, deg	2.9–50.0	5.3–50.5	3.5–50.5	3.3–50.5
μ, mm^{-1}	2.004	2.998	0.934	0.572
$F(000)$	960	872	896	1664
No. of obsd reflns	7460	1936	8695	7596
No. of params refnd	478	109	487	514
Goodness of fit	1.095	1.144	1.018	1.051
R1	0.041	0.055	0.033	0.031
wR2	0.098	0.149	0.084	0.081

Z. Qiu, *Late Transition Metal-Carboryne Complexes*, Springer Theses,
DOI: 10.1007/978-3-642-24361-5, © Springer-Verlag Berlin Heidelberg 2012

	II-6	III-3c	III-3e	III-3g
Formular	$C_{16}H_{44}B_{10}NiP_2$	$C_{11}H_{17}B_{10}F_3$	$C_{13}H_{24}B_{10}O_3$	$C_{16}H_{22}B_{10}$
Crystal size (mm)	$0.40 \times 0.30 \times 0.30$	$0.50 \times 0.40 \times 0.30$	$0.40 \times 0.30 \times 0.20$	$0.50 \times 0.40 \times 0.30$
fw	465.26	314.35	336.42	322.44
Crystal system	Monoclinic	Monoclinic	Monoclinic	Monoclinic
Space group	$P2_1/c$	$P2_1/c$	$P2_1$	$P2_1/c$
a, Å	10.290(1)	12.322(2)	7.275(1)	12.518(1)
b, Å	25.985(3)	18.702(2)	10.326(1)	7.549(1)
c, Å	11.492(1)	7.547(1)	13.105(1)	20.712(2)
α, deg	90	90	90	90
β, deg	110.827(2)	105.034(3)	103.601(2)	97.800(2)
γ, deg	90	90	90	90
V, Å^3	2872.2(5)	1679.5(4)	956.9(2)	1938.9(3)
Z	4	4	2	4
D_{calcd}, Mg/m^3	1.076	1.243	1.168	1.105
Radiation (λ), Å	0.71073	0.71073	0.71073	0.71073
2θ range, deg	3.1–50.0	3.4–56.0	5.1–50.0	3.3–56.0
μ, mm^{-1}	0.789	0.085	0.069	0.054
$F(000)$	992	640	352	672
No. of obsd reflns	5059	4031	3224	4655
No. of params refnd	262	217	235	235
Goodness of fit	1.838	1.032	1.036	1.005
R1	0.153	0.072	0.043	0.063
wR2	0.426	0.198	0.108	0.157

	III-5q	III-7r	III-8r	III-9r
Formular	$C_6H_{18}B_{10}O_2$	$C_{16}H_{26}B_{10}N_2$	$C_{16}H_{24}B_{10}N_2$	$C_{39}H_{81}B_{30}ClLiN_3Ni_3O_6$
Crystal size (mm)	$0.40 \times 0.30 \times 0.20$	$0.50 \times 0.40 \times 0.30$	$0.40 \times 0.30 \times 0.20$	$0.50 \times 0.40 \times 0.30$
fw	230.30	354.49	352.47	1230.89
Crystal system	Monoclinic	Orthorhombic	Monoclinic	Monoclinic
Space group	$P2_1/n$	$Pnma$	$C2/c$	$P2_1/n$
a, Å	21.807(4)	10.636(2)	19.594(1)	15.252(3)
b, Å	10.773(2)	19.055(4)	10.876(1)	23.471(4)
c, Å	24.597(5)	10.113(2)	19.714(1)	18.257(4)
α, deg	90	90	90	90
β, deg	108.21 (1)	90	100.018(1)	91.51(1)
γ, deg	90	90	90	90
V, Å^3	5489(2)	2050(1)	4137(1)	6533(2)
Z	16	4	8	4
D_{calcd}, Mg/m^3	1.115	1.149	1.132	1.251
Radiation (λ), Å	0.71073	0.71073	0.71073	0.71073
2θ range, deg	2.1–50.0	4.3–56.1	4.2–55.6	2.8–56.0
μ, mm^{-1}	0.062	0.060	0.059	0.939
$F(000)$	1920	744	1472	2552
No. of obsd reflns	9652	2549	4841	15772
No. of params refnd	649	133	253	748

(continued)

Appendix

(continued)

	III-5q	III-7r	III-8r	III-9r
Goodness of fit	1.079	1.023	1.036	1.010
R1	0.076	0.069	0.062	0.062
wR2	0.217	0.181	0.174	0.146

	IV-1d	IV-1i	IV-5e	IV-5l
Formular	$C_{18}H_{25}B_{10}N$	$C_{12}H_{26}B_{10}O_2$	$C_{23}H_{17}N$	$C_{20}H_{20}O_3$
Crystal size (mm)	$0.50 \times 0.40 \times 0.30$	$0.40 \times 0.30 \times 0.20$	$0.40 \times 0.30 \times 0.20$	$0.50 \times 0.40 \times 0.30$
fw	363.49	310.43	307.38	308.36
Crystal system	Monoclinic	Monoclinic	Monoclinic	Monoclinic
Space group	$P2_1/c$	$P2_1$	$P2_1/c$	$P2_1/n$
a, Å	12.961(2)	10.412(1)	11.313(2)	12.573(3)
b, Å	7.054(1)	14.952(2)	15.952(3)	9.332(2)
c, Å	23.025(3)	12.096(1)	9.467(2)	13.657(3)
α, deg	90	90	90	90
β, deg	98.98(1)	102.44(1)	98.285(4)	90.587(4)
γ, deg	90	90	90	90
V, Å^3	2079.1(5)	1839.0(4)	1722.6(5)	1602.4(6)
Z	4	4	4	4
D_{calcd}, Mg/m^3	1.161	1.121	1.185	1.278
radiation (λ), Å	0.71073	0.71073	0.71073	0.71073
2θ range, deg	3.2–56.0	3.4–56.0	3.6–56.1	4.4–56.0
μ, mm^{-1}	0.059	0.062	0.068	0.085
$F(000)$	760	656	648	656
No. of obsd reflns	5023	8294	4147	3861
No. of params refnd	262	433	217	208
Goodness of fit	1.023	1.005	0.972	1.027
R1	0.057	0.059	0.051	0.054
wR2	0.158	0.122	0.129	0.137

	V-1h	V-1n	V-1o	V-1'o
Formular	$C_{14}H_{30}B_{10}$	$C_{34}H_{30}B_{10}$	$C_{22}H_{30}B_{10}O_2$	$C_{22}H_{30}B_{10}O_2$
Crystal size (mm)	$0.40 \times 0.30 \times 0.20$	$0.40 \times 0.30 \times 0.20$	$0.40 \times 0.30 \times 0.20$	$0.30 \times 0.20 \times 0.20$
fw	306.48	546.68	434.56	434.56
Crystal system	Monoclinic	Triclinic	Triclinic	Triclinic
Space group	$P2_1/m$	$P(-1)$	$P(-1)$	$P(-1)$
a, Å	9.932(1)	11.171(2)	6.546(6)	8.871(1)
b, Å	10.652(1)	11.415(2)	12.647(11)	9.580(1)
c, Å	10.052(1)	13.505(3)	15.303(14)	15.712(1)
α, deg	90	94.33(1)	93.46(2)	80.99(1)

(continued)

(continued)

	V-1h	V-1n	V-1o	V-1'o
β, deg	115.14(1)	106.82(1)	90.41(2)	88.55(1)
γ, deg	90	99.08 (1)	100.42(2)	71.30(1)
V, Å^3	962.6(1)	1614.3(5)	1243.5(19)	1248.7(2)
Z	2	2	2	4
D_{calcd}, Mg/m^3	1.057	1.125	1.161	1.156
radiation (λ), Å	0.71073	0.71073	0.71073	1.54178
2θ range, deg	4.5–50.0	3.2–50.0	2.7–50.5	5.7–135.4
μ, mm^{-1}	0.051	0.059	0.065	0.484
$F(000)$	328	568	456	456
No. of obsd reflns	1787	5668	4079	4282
No. of params refnd	127	397	307	308
Goodness of fit	1.020	1.045	1.119	0.942
R1	0.069	0.059	0.113	0.076
wR2	0.192	0.151	0.307	0.176

	V-6o	V-8b	V-9	VI-4a
Formular	$C_{12}H_{22}B_{10}O$	$C_{12}H_{24}B_{10}$	$C_{46}H_{86}B_{20}Cl$ $LiN_2Ni_2O_5$	$C_{14}H_{30}B_{10}$
Crystal size (mm)	$0.40 \times 0.30 \times 0.20$	$0.50 \times 0.40 \times 0.30$	$0.40 \times 0.30 \times 0.20$	$0.50 \times 0.40 \times 0.30$
fw	290.40	276.41	1123.18	306.48
Crystal system	Triclinic	Orthorhombic	Triclinic	Monoclinic
space group	$P(-1)$	$Pbca$	$P(-1)$	$P2_1/c$
a, Å	7.807(8)	18.004(3)	11.388(3)	9.63(1)
b, Å	9.509(1)	18.596(3)	14.951(3)	17.65(1)
c, Å	12.428(1)	19.838(3)	19.334(4)	12.05(1)
α, deg	73.32(1)	90	86.53(1)	90
β, deg	84.61(1)	90	83.80(1)	111.74(1)
γ, deg	77.33(1)	90	74.03(1)	90
V, Å^3	861.8(1)	6639(2)	3145(1)	1901(2)
Z	2	16	2	4
D_{calcd}, Mg/m^3	1.119	1.106	1.186	1.071
Radiation (λ), Å	0.71073	0.71073	0.71073	0.71073
2θ range, deg	3.4–50.5	3.8–50.0	2.1–50.0	4.3–50.0
μ, mm^{-1}	0.058	0.053	0.683	0.052
$F(000)$	304	2336	1184	656
No. of obsd reflns	3107	5845	11023	3355
No. of params refnd	208	397	694	217
Goodness of fit	1.014	1.044	1.025	1.044
R1	0.056	0.072	0.080	0.075
wR2	0.143	0.184	0.205	0.161

Appendix

	VI-4b	VI-4d	VI-4j	VI-4m
Formular	$C_{15}H_{32}B_{10}$	$C_{17}H_{38}B_{10}Si$	$C_{31}H_{32}B_{10}$	$C_{23}H_{32}B_{10}$
Crystal size (mm)	$0.40 \times 0.30 \times 0.20$	$0.50 \times 0.40 \times 0.30$	$0.50 \times 0.40 \times 0.30$	$0.40 \times 0.30 \times 0.20$
fw	320.51	378.66	512.67	416.59
Crystal system	Orthorhombic	Orthorhombic	Trigonal	Monoclinic
Space group	$Pna2_1$	$P2_12_12_1$	$R(-3)$	$P2_1$
a, Å	17.16(2)	9.771(3)	38.990(1)	12.641(1)
b, Å	9.76(1)	14.550(5)	38.990(1)	8.951(1)
c, Å	12.10(1)	16.814(6)	12.161(1)	22.288(2)
α, deg	90	90	90	90
β, deg	90	90	90	98.53(1)
γ, deg	90	90	120	90
V, Å^3	2026(3)	2390(1)	16011(1)	2493.8(4)
Z	4	4	18	4
D_{calcd}, Mg/m^3	1.051	1.052	0.957	1.110
Radiation (λ), Å	0.71073	0.71073	0.71073	0.71073
2θ range, deg	4.7 to 50.0	3.7 to 50.5	3.5 to 50.0	1.8 to 50.0
μ, mm^{-1}	0.051	0.100	0.050	0.056
$F(000)$	688	816	4824	880
No. of obsd reflns	2873	4328	6196	8299
No. of params refnd	226	253	371	595
Goodness of fit	1.082	1.049	1.056	1.001
R1	0.048	0.039	0.069	0.052
wR2	0.106	0.108	0.205	0.103

	VI-5m	VI-7b
Formular	$C_{23}H_{32}B_{10}$	$C_{13}H_{22}B_{10}$
Crystal size (mm)	$0.50 \times 0.40 \times 0.30$	$0.50 \times 0.40 \times 0.30$
fw	416.59	286.41
Crystal system	Monoclinic	Monoclinic
Space group	$P2_1/n$	$P2_1/c$
a, Å	8.939(1)	9.31(1)
b, Å	17.671(2)	7.73(1)
c, Å	16.068(2)	24.42(2)
α, deg	90	90
β, deg	91.45(1)	91.05(1)
γ, deg	90	90
V, Å^3	2537.3(5)	1757(2)
Z	4	4
D_{calcd}, Mg/m^3	1.091	1.083
Radiation (λ), Å	0.71073	0.71073
2θ range, deg	3.4–50.5	3.3–50.0
μ, mm^{-1}	0.055	0.052
$F(000)$	880	600
No. of obsd reflns	4575	3092
No. of params refnd	298	208
Goodness of fit	1.094	1.035
R1	0.082	0.090
wR2	0.248	0.251